PHOTOELECTROCHEMICAL WATER SPLITTING

Photoelectrochemical Water Splitting

P. Pérez Rodríguez

ARCLER PRESS

www.arclerpress.com

Photoelectrochemical Water Splitting

P. Pérez Rodríguez

Arcler Press
2010 Winston Park Drive,
2nd Floor
Oakville, ON L6H 5R7
Canada
www.arclerpress.com
Tel: 001-289-291-7705
001-905-616-2116
Fax: 001-289-291-7601
Email: orders@arclereducation.com

ISBN: 978-1-77361-095-5 (Hardcover)

About the Author

Paula received her Bachelor degree in Chemical Engineering in 2011 from the Universidad de Cantabria in Spain, with an exchange program in Oregon State University in 2009-2010. She obtained her M.Sc. in Sustainable energy technology in 2013 in TU Delft (Netherlands). Her M.Sc. thesis in the PVMD group dealt with Photoelectrochemcial devices for solar water splitting. Since May 2014 she is a PhD student working on the development and optimization of the monolithic photovoltaic (PV)/ photoelectrochemical (PEC) devices based on earth abundant materials such as silicon and carbon with ground-breaking high solar-to-hydrogen (STH) conversion efficiency for water splitting.

Contents

List of Figures

List of Tables

Table 16. Main assumed parameters to calculate the costs of the overall plant design. From (Pinaud, et al., 2013)

Nomenclature

a_i	Activity of species i
A_w	Atomic weight
ABPE	Applied bias photon-to-current efficiency
ALD	Atomic layer deposition
ALE	Atomic layer epitaxy
AM	Air mass
APCE	Absorbed photon-to-current efficiency
BPM	Bipolar membrane
c	Speed of light at vacuum
C_d	Differential capacitance
C_H	Helmholtz layer capacitance
$C_{H,d}$	Helmholtz layer differential capacitance
c_i	Concentration of species i
C_{SC}	Space charge capacitance
CAES	Compressed air energy storage
CB	Conduction band
CBD	Colloidal bath deposition
CE	Counter electrode
CVD	Chemical vapor deposition
D_n	Diffusion coefficient of electrons
D_p	Diffusion coefficient of holes
DC	Direct current
DOS	Density of states
DSSC	Dye sensitize solar cells
e	elemental charge
E	Photon energy
$\bar{E}$	Electric field
E^0	Standard potential for the overall redox reaction

E_C	Minimum conduction band energy
E_{decomp}	Decomposition reaction potential
E_{fb}	Flat band potential
E_{Fi}	Fermi energy level
E_g	Bandgap energy
E_{onset}	Onset potential
E^0_{ox}	Standard potential for the oxidation half-reaction
E^0_{red}	Standard potential for the reduction half-reaction
E_{redox}	Redox reaction potential
E_V	Maximum valence band energy
EIS	Electrochemical impedance spectroscopy
EQE	External quantum efficiency
EPD	Electrophoretic deposition
f	Activity coefficient
F	Faraday constant
FTO	Fluor doped tin oxide
G	Gibbs free energy
GC	Gas chromatography
GHG	Greenhouse gas
h	Planck's constant
HDPE	High density polyethylene
HER	Hydrogen evolution reaction
i_0	Exchange current
I	Electric current
$I(\lambda)$	Transmitted radiation
$I_0(\lambda)$	Incident radiation
IPCE	Incident Photon-to-current efficiency
IR	Infrared
IS	Integrating sphere
ITO	Indium doped tin oxide
J	Current density
J_{br}	Current density associated with radiative and non-radiative recombination in the semiconductor bulk
J_{dr}	Current density associated with surface recombination due to defects
J_{et}	Electron transfer current density associated with majority charge

	carriers traversing the interfacial barrier
J_{ph}	Photocurrent density
J_{SC}	Short circuit current
J_{ss}	Current density associated with the surface recombination due to defects
J_t	Current density associated with the tunneling process
k	Boltzmann constant
$k(\lambda)$	Wavelength dependent absorption coefficient
L	Absorption path-length
LPCVD	Low-pressure chemical vapor deposition
m	Mass
MBE	Molecular beam epitaxy
MOCVD	Metal-organic chemical vapor deposition
MOD	Metal organic deposition
M-S	Mott-Schottky
MS	Mass spectrometer
n	Amount of electrons in the material
N_C	Effective density of charge carriers in the conduction band
$\dot{n}_{H_2}$	Molar flow of hydrogen
n_i	Number of moles of species i in a particular phase α
N_V	Effective density of charge carriers in the valence band
NHE	Normal hydrogen electrode
NW	Nanowires
OCP	Open-circuit potential
OER	Oxygen evolution reaction
OP_{HER}	Overpotential associated with the hydrogen evolution reaction
OP_{OER}	Overpotential associated with the oxygen evolution reaction
p	Amount of holes in the material
P_i	Electrochemical potential
P_{total}	Total irradiation per area
PEC	Photoelectochemical
PECVD	Plasma-enhanced chemical vapor deposition
PHS	Pumped hydro storage
PMMA	Poly-methyl metacrylate
PSA	Pressure swing absorption

PVD	Physical vapor deposition
q	Elementary charge
Q_{SS}	Charge accumulation at the surface
QD	Quantum dot
R	Ideal gas constant
RBD	Reactive ballistic deposition
RE	Reference electrode
RF	Radio frequency
RGO	Reducing graphene oxide
RMS	Reactive magnetron sputtering
SCE	Saturated calomel electrode
SEM	Scanning electron microscopy
SHE	Standard hydrogen electrode
SP	Spray pyrolysis
SRH	Shockley-Read-Hall recombination
STF	Sun-to-fuel
STH	Solar-to-hydrogen
t	Time
T	Temperature
TCO	Transparent conductive oxide
TSA	Temperature swing absorption
U	Potential
U^0_{redox}	Standard redox potential
USP	Ultrasonic spray pyrolysis
UV/Vis	Ultraviolet – visible
V	Voltage
V_b	Applied bias voltage
VHF	Very high frequency
VB	Valence band
WE	Working electrode
XPS	X-ray photoelectron spectroscopy
XRD	X-ray diffraction
z_i	Charge number of species i
α_a	Transfer coefficient for the anodic reaction
ΔH	Enthalpy change of a reaction

λ	Wavelength
η	Efficiency
η_F	Faradaic efficiency
μ	Mobility
μ_i	Standard chemical potential
θ	Angle of incidence
φ	Electrical potential
ν_i	Coefficient to indicate product (positive) or reactant (negative)
∇_n	Concentration gradient of electrons
∇_p	Concentration gradient of holes

Preface

The world energy demand has been steadily growing in the past decades as the world population increases and more nations develop to higher standards of living. Traditional solutions such as fossil fuels and nuclear energy have not been able to arrive to a sustainable plan on how to supply this energy the next few decades, and many armed conflicts have been started due to the limited access of such a scarce resource. Moreover, they are responsible for toxic waste and greenhouse gas emissions, which are causing one of the most important environmental crises in the history of the planet. Thus, alternative solutions must be considered in the energy transition that would be able to supply the needed energy in the future. Renewable energies, including wind, solar, biomass, wave and geothermal among others, are the main hope to cover the energy needs of society in the future, since it is a more sustainable way of harvesting energy and these resources are virtually infinite in terms of time scalability. In particular, solar energy is the most readily abundant energy source in most areas of the world, since the amount of solar energy received by the earth every year is thousands of times higher than the energy demand. In addition, it is considered one of the sources with the least impact in the surrounding environment among all the renewable energy sources, since it does not produce sound, and the most common techniques do not produce toxic waste. For these reasons, solar energy has experimented a steep growth in production and implementation recently.

However, if solar energy sources are to play a crucial role in the necessary energy transition, they must be able to supply a constant amount of power throughout the year. One of the main problems that solar energy faces is its daily and seasonal fluctuations due to the nature of this source, which threaten to destabilize the electricity network if solar energy is to be installed at very large scale. Thus, reliable systems for energy storage must be installed to assure that the fluctuations in the energy source do not affect the energy supply chain. So far, batteries have been used as the main energy storage system. However, they are rather bulky and expensive, with toxic and rare materials at their core, and thus ineffective for long-term energy storage. One of the most promising approaches to this issue, especially to long term storage, is the use of hydrogen as an energy storage material for solar energy. Hydrogen has a high energy density and can be stored as a pressurized gas, a liquid, a metal hydride, or further converted in more common hydrocarbons such as methane or ethanol.

An interesting way to achieve hydrogen using solar energy is to drive a photoelectrochemical (PEC) reaction, in which a semiconductor material is excited, producing an electron-hole pare that would be directly used to drive the electrochemical reaction of water electrolysis, also called water splitting. This book gives an account of the main physical principles governing this process, identifying important barriers and areas of potential improvements. In particular, there seems to be three major steps that may limit the performance of these devices: the charge carrier separation in the semiconductor material used as photoelectrode; the interface between the semiconductor and the electrolyte, including the charge injection from one to the other, the catalytic activity at the surface and the possible stability issues that can occur; and the ion transfer and optimum pH within the electrolyte itself. All these issues have been further explored here.

The main strategies applied so far to achieve a good charge carrier generation, separation and injection are reviewed within this book, with the most important materials investigated in the field to date. There seems to be a special focus historically in TiO2 and Fe2O3, as they are among the first materials to be investigated and developed. Here, the main reasons behind these choices were investigated, especially based on the physical principles previously explained. In addition, it is also interesting to look at possible catalysts for these reactions, both in the areas of precious metals and earth abundant materials, and to further explore the strategy of including protective layers to avoid corrosion of the photoelectrodes. Moreover, some emerging trends such as new more complex materials, nanostructures of such semiconductors, and the application of multijunctions and membranes are reviewed.

In addition, the fabrication techniques and measuring methods are listed, identifying possible sources of practical challenges. Practical issues regarding the fabrication techniques seem to have been one of the main limits for the performance of more earth-abundant materials, and thus further understanding on how these techniques affect the material properties of the semiconductors fabricated up to date. Moreover, there has been several instances of irregular or uninformed reporting of performances within this field, thus, understanding the different measurement techniques and how to relate those to the final expected performance and calculated solar-to-hydrogen efficiencies is crucial to raise the reporting standards of the field.

Finally, the economic feasibility of such approach into a reactor design and a hydrogen production plant are discussed, allowing to draw some general conclusions and indicating future approaches that must be thoroughly investigated and improve to arrive to an economic and efficient PEC system. This is especially relevant since, so far, most of the PEC devices reported are in the scale of millimeters to centimeters. Thus, looking forward to the implementation of such devices at large scale, possible bottlenecks and additional equipment needed is of vital importance for a reliable economic analysis.

In summary, this book tries to give an overview of the field of photoelectrochemical water splitting, by looking at the physics, the state-of-the-art devices, emerging trends and fabrication and measurement techniques. Moreover, the economic feasibility based on these reported performances and trends has been investigated. This analysis allows drawing some conclusions in the feasibility of the methods presented, and their role on the energy transition for future societies.

Author

CHAPTER 1

INTRODUCTION

One of the main challenges facing the world today is to assure the energy supply to cover the world's energy demands. In 2012, the world energy demand was 160,000 TWh, and it is expected to grow to 180,000 TWh in the next 30 years. (EIA, 2016) There are two main reasons for this increase: the increasing world population and the development of several countries, raising the standards of living and the need of energy. (EIA, 2016) To overcome this challenge, renewable energies such as solar energy need to play an important role in the energy market. In particular, solar energy has a great potential, since it is the most abundant renewable energy source. (Edenhofer, et al., 2012) However, this process is not only about the energy production, but also about energy storage due to the intermittent nature of many renewable energy sources. In particular, in certain climates the seasonal variations of solar energy can be a challenge regarding storage and stable energy supply. One of the most promising ways to tackle this problem is the use of solar fuels, namely chemical compounds produced by using solar energy that can release energy at the desired times. In order to better understand how these variations, the possible storage strategies of solar energy and the need of new solutions, this section explores the energy current scenario, and some of the solutions proposed would be explained, in particular the meaning and importance of solar fuels.

THE ENERGY SCENARIO

The energy demand is associated with social and economic development, which is following an increasing trend at the moment. Some of the most basic tasks that energy covers are lighting, cooking, mobility or commutation. (Dincer, 2000) These functions are mainly powered either by an electrical supply or directly by fuel combustion. Both of these supplies are often dominated by fossil fuels such as coal, oil or gas. (Uno, 1995) However, these supplies lead to carbon dioxide (CO_2) and other greenhouse gases (GHGs) emissions. The increase of GHGs emissions in the atmosphere leads to an average temperature rise, also known climate change. (Edenhofer, et al., 2012)

In order to reduce the GHGs emissions, the use of fossil fuels needs to be reduced. This can be done by a series of strategies, including energy conservation, efficiency, nuclear energy or renewable energy sources. Renewable energies in particular have other benefits such as contributing to social and economic development, energy access, or energy security. However, it also has disadvantages such as the high investment costs associated, which often translates in a need for incentives at the beginning to stimulate its inclusion in the energy mix. (Menyah & Wolde-Rufael, 2010)

The energy sources considered to be renewable sources include solar energy, wind, tidal energy, geothermal energy biomass and hydropower. However, not all these technologies are equally suitable. The preferred renewable energy source applied would depend on the availability of such source in a specific location and on the development stage of that technology in particular. (Mardani, et al., 2015) In addition, although all the mentioned sources are renewable energies, not all of them are assumed to be sustainable, with the controversial cases of biomass and hydropower. (Kaunda, et al., 2012) Based on these parameters, solar energy seems to be one of the most suitable technologies for most locations. It is vastly abundant worldwide, as shown in **Figure** 1, and the Technologies show a high degree of maturity. (Ellabban, et al., 2014) Although not in such a large scale, wind energy is also expected to play an important role in this implementation.

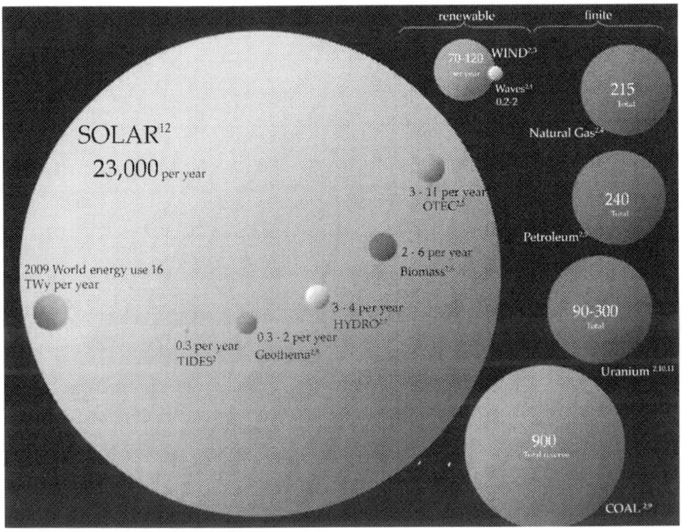

Figure 1: Estimated finite and renewable planetary energy reserves (TW-year). For the finite resources, the total recoverable resources. For the renewable resources, it is the yearly potential. From (IEA, 2009)

In fact, although the scales compared to fossil fuels are still small, the installation of renewable and sustainable energy sources has been increasing in the last years. Currently, renewable energy represent about 13 % of the current energy supply. (Edenhofer, et al., 2012) As **Figure** 2 illustrates, it is apparent that, although hydropower still remains the biggest renewable energy installed in the world, its use is relatively stable. Alternatively, other more sustainable energy sources have grown radically in the past years, with the percentage change of wind energy installed from 2006 to 2015 of 487.7 % and a percentage of 3404.9 % for solar energy. Note that this only applies to the electricity production. If the complete energy use is considered, bioenergy also plays a crucial role, especially in the transportation system transition. (Edenhofer, et al., 2012)

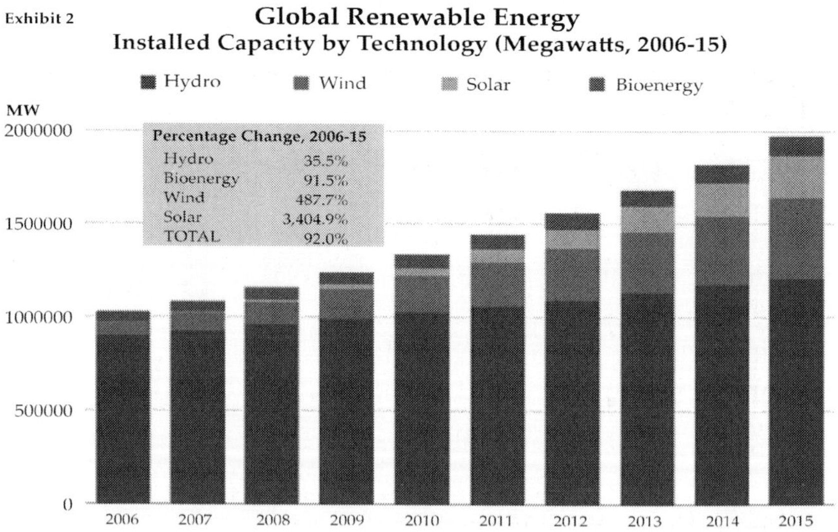

Figure 2: Global renewable energy installed capacity by technology, from 2006 to 2015. From (Whiteman, et al., 2016)

It is also important to notice that the leading investors in renewable energy do not only include developed countries, but also developing areas, as shown in Table 1. This shows that, although the benefits of renewable energy are not immediate and the investment costs are relatively high, the benefits related to energy security and lower pollution levels outperform the difficulties. The recent Paris agreement regarding climate change and GHGs emission has further eased this transition by placing strategies and target for both developed and developing countries. (Rogelj, et al., 2016)

Table 1: Leading investor countries in renewable energy. From (McCrone, et al., 206)

Rank	Country	Investment in 2015 (USD Billion)
1	China	102.9
2	U.S.	44.1
3	Japan	36.2
4	U.K	22.2
5	India	10.2

6	Germany	8.5
7	Brazil	7.1
8	South Africa	4.5
9	Mexico	4.0
10	Chile	3.4

One last remark concerns the systems that would be needed surrounding the renewable energy source, such as a smart and flexible grid that can cope with such an energy source, or the needed storage, conversion and transportation systems. (Dunn, et al., 2011)In order to target some of these challenges for the case of solar energy, the concept of energy storage would be further explored.

ENERGY STORAGE METHODS

As already stated, energy storage is vital for integrating renewable energies within our current energy system. In particular, solar energy has both daily and seasonal fluctuation inherent to it that need to be covered. This section reviews the different technological options available for covering for these fluctuations.

Batteries are considered at the moment the most promising energy storage method in the market. However, there are other methods worth considering, such as flywheels, supercapacitors, pumped hydro storage (PHS), fuel cells, or compressed air energy storage (CAES). **Figure** 3 shows the potential for autonomy time and power rate of each available technique. From this graph, it is evident that methods such as flywheels and supercapacitors are only useful to stabilize the electrical grid in a timescale from seconds to minutes, due mainly to spikes in consumption or transient clouds over a solar system. Most batteries fall under the category bridging power, in the scale of hours to days. In the case of solar energy, they could be considered for daily variations between day and night. However, they are often considered too bulky, expensive and employing compounds too expensive to scale such a system to bigger sizes.

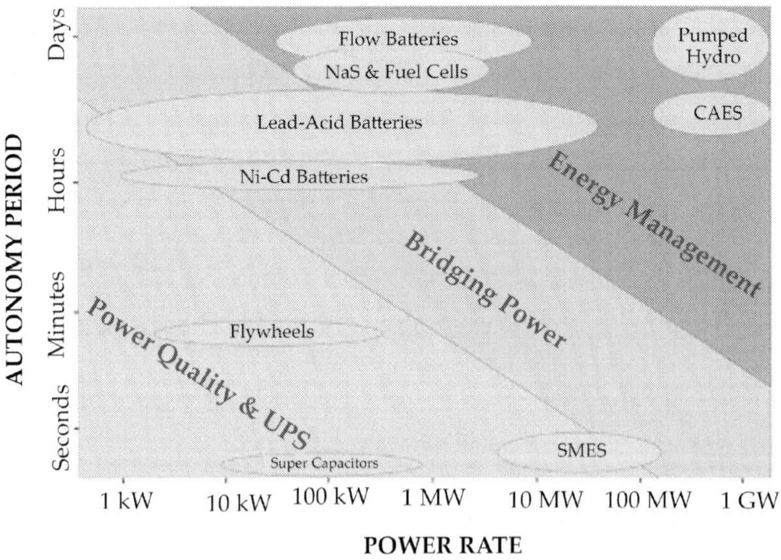

Figure 3: Energy storage systems applications' range, presenting the autonomy period and power covered by each technology. Form (Kaldellis & Zafirakis, 2007)

Thus, from this analysis, it seems that there are only a few options for seasonal storage of energy: PHS, CAES, fuel cells and flow batteries. To decide the most suitable solution, (Yoon, et al., 2016) discuss not only the autonomy, but also the energy and power density of these methods. That represents the amount of enrgy that can be stored in each technology per kilogram of material. Looking at this **Figure**, it is evident that natural gas, and in general fossil fuels, are the most energy dense materials, and among the ones with a higher rated power. However, as already mentioned, fossil fuels are causing numerous environmental and political issues in our current world, and thus need to be replaced. Considering that, althoug PHS and CAES seem like a mature and feasible option for long term storage (Chen, et al., 2009), they also have a relatively low energy density, meaning that a very large facility is needed to accommodate the necessary energy storage for seasonal variations. Thus, the only remaining option for this timescale with sustainable results would be hydrogen storage. Hydrogen has quite a large energy density, and can be efficiently converted back to electrical energy by using fuel cells, making it the most promising candidate among all the reviewed ones. Moreover, hydrogen can be further combined with CO_2 and other carbon compounds to synthesize more traditional fuels, which would then create an ideally close cycle of energy, in which CO_2 would be produced, but also reduced by its use in further fuel synthesis (Olah, 2005) (Roy, et al., 2010).

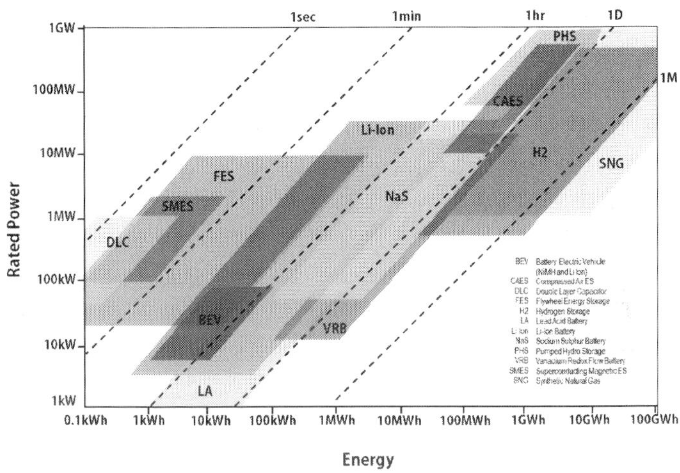

Figure 4: Comparison of power density and energy density of different energy storage technologies. From (Yoon, et al., 2016)

However, the conversion from solar energy into hydrogen production must then be considered. There are several routes for this process with different degrees of success and efficiencies. This book, and in particular the following sections, present a general idea of the reactions and routes involved into this process, and possible areas for further improvements to arise.

SOLAR FUELS

As previously mentioned, solar energy is bound to have a crucial role on future energy scenarios. However, solar energy is a fluctuating source, with both daily variations and seasonal variations. These fluctuations can severely affect the electricity networks and electricity security. Thus, storage is a crucial aspect to implement solar energy in a large scale. Among the storage possibilities, solar fuels offer a safe and efficient way of storing energy. These are essentially chemical fuels, such as hydrogen, methane, methanol, or gasoline. Generally, they have a high energy density, they are cheap and, in most cases, compatible with the current energy infrastructure. Moreover, the energy in this case is relatively easy to transport, unlike other energy storage technologies such as batteries or flywheels. (Bossel, et al., 2003)

An interesting and efficient route to achieve hydrogen production using solar energy is by applying photoelectrochemistry (PEC) in order to split the water molecule into hydrogen and oxygen. Visible light holds enough energy to chemically split the water molecule into hydrogen and oxygen. Hydrogen can then be used as a fuel in itself or can be further used to create other fuels when combined with CO_2, which can be extracted from the atmosphere or recovered from cars, combustion plants and other concentrated CO_2 sources. (Izumi, 2013)

The basic water splitting reaction to create hydrogen gas can be described as

$$H_2O \rightarrow H_2 + \frac{1}{2} O_2 \qquad \Delta G = 237 \text{ kJ/mol} \qquad \text{Equation 1}$$

Hydrogen use as a fuel from water splitting has several advantages, since water is readily available and the purity of the resulting product allows for the use of highly efficient fuel cells in order to be converted back into electricity. (Ahmed & M.Krumpelt, 2001) However, there are also disadvantages attached to the use of hydrogen as a fuel. Firstly, it is a very small and highly reactive gas in the presence of oxygen, introducing potential hazards in the system. In addition, it is a challenge to efficiently store hydrogen due to its low volumetric energy density. Strategies to solve this issue include high pressure containers, liquid storage at very low temperatures or physisorption into metal halides. (Ni, 2006) Finally, hydrogen is able to weaken the steel piping that is currently being used for fossil fuel transportation, meaning that existing infrastructure could not be used in the presence of hydrogen, and thus all infrastructures would need to be renewed to implement a hydrogen economy. (Melania, et al., 2013)

Thus, due to these limitations in hydrogen storage and transportation, other possible fuels have been explored. The most straight-forward method would be to implement the Fischer-Tropsch reaction to further convert the hydrogen into other carbon based fuels. (Adesina, 1996) First, CO is needed as an input, which can be obtained as

$CO_2 + H_2 \rightarrow H_2O + CO \qquad \Delta H = 42\ kJ/mol$ **Equation 2**

Then, the basic Fischer-Tropsch reaction can be implemented as (Dry, 2002)

$(2n+1)\ H_2 + n\ CO \rightarrow C_nH_{2n+2} + n\ H_2O$ **Equation 3**

where n defines the hydrocarbon obtained.

This method is very well known in industry, and thus the know-how is very high. However, it is expected that the resulting mixture, although predominant in one specific hydrocarbon, would present a mixture of several possible hydrocarbon chains. Other methods, such as direct CO_2 reduction also exists, with the disadvantage of low maturity levels, especially for its application in industry. (Schulz, 1999) Thus, for the purpose of this book, only hydrogen is going to be considered as the first step towards solar fuels.

ROUTES TO SOLAR FUEL PRODUCTION

There are several routes that can be considered to produce solar fuels: photoelectrochemical (PEC) water splitting, photocatalytic water splitting, indirect PV/electrolyzer systems, thermochemcial conversion, photobiological methods, molecular artificial photosynthesis, plasma-enhanced conversion, or mechano-catalytic methods. (Krol & Gratzel, 2012) This book mainly focuses on the first method, PEC, although it would briefly touch upon other methods such as PV/electrolyzer systems. The PEC method has the advantage that hydrogen and oxygen are produced at different electrodes, and thus the gas separation process is facilitated and possible hazardous situations are minimized. In addition, this method can be applied at room temperature, which translates in an easier, cheaper and safer process. Finally, these devices use completely inorganic materials, often abundant and cost effective, increases the robustness and durability of the systems.

Although these advantages can also apply to PV/electrolizer systems, which have been constructed commercially with an efficiency of about 12 %, (Gibson & Kelly, 2008) the efficiencies of PEC devices are expected to be potentially higher due to reduced losses on the cables and more accurate optimization of the whole device for water splitting. In addition, a PEC approach could reduce the costs related to packaging and installation, since the overall system would be a much simpler configuration. (Krol & Gratzel, 2012)

PEC water splitting was first demonstrated by Fujishima and Honda in 1972 (Fujishima & Honda, 1972). Since then, numerous developments have taken place in order to improve the performance of these PEC devices, with Khaselev and Turner reaching a PEC solar-to-hydrogen efficiency of 12.4 % already in 1998 (Khaselev & Turner, 1998), highlighting the potential of this technology. The main idea of this method is that, when a semiconductor device is immersed in an aqueous electrolyte and illuminated with sunlight, an electron-hole pair would be created and collected at the semiconductor/electrolyte interfaces, being able to drive the solar water splitting reaction.

Although the initial idea is rather simple, there are many factors involved in this process that would need to be taken into account in order to reach efficient conversion from light into chemical energy. First, the reaction occurring at the semiconductor/electrolyte interface is an electrochemical reaction, and thus it has a voltage associated to it that needs to be covered by the semiconductor produced voltage. Moreover, this semiconductor must also be able to efficiently absorb light and separate the charge carriers that would drive the chemical reaction. In addition, a catalyst is introduced at the surface to facilitate the carrier injection from the semiconductor into the electrolyte. The electrolyte in itself also plays a crucial role in the reaction, and its conductivity, pH and temperature should be controlled. Moreover, not only should the device be efficient, but also stable at the working conditions and based on cheap and abundant materials. Finally, to arrive to an economically feasible device not only must the device be efficient and cheap, but it must also be stable during a long period of time, even when exposed to harsh conditions such as extreme pH. (Chen, et al., 2013) In the course of this book, all these elements would be discussed and possible strategies would be proposed. In addition, an economic analysis of the different approaches, as well as an analysis of further optimization during the scale-up process would be given.

SUMMARY

The energy demand is increasing due to the increase in world population and the development in many countries. Fossil fuels, apart from polluting the environment, would also not be able to cope with this increase in demand in the long term. Thus, a transition in the energy scenario is necessary to more sustainable sources such as solar energy. The main issue with solar energy is the need for short-term and long-term storage. Batteries are usually considered suitable for short-term storage, but other technologies are needed for long-term storage. One of the most promising methods is by converting solar energy into hydrogen for storage either as gas or liquid, or to further converts it into other fuels such as methane or ethanol.

To arrive to these chemical fuels from solar energy, several routes can be taken. This book concentrates mainly in photoelectrochemical (PEC) techniques due to its potential to efficiently produce hydrogen using relatively simple techniques. In this method, light excites a semiconductor, generating electron-hole pairs, which would be separated and collected. Then, they would be injected into an electrolyte, where they would drive the electrolysis reaction of splitting water. Several factors must be consider to optimize a PEC device, such as the semiconductor itself, its structure, the electrolyte and its effect on the semiconductor, or the possible catalyst. Further chapters will deal with these characteristics and further analysis on the deposition and measurement techniques, as well as the economic factors of PEC water splitting.

CHAPTER 2

PHOTOELECTROCHEMICAL (PEC) PRINCIPLES

In order to understand the requirements and improvement strategies applied in PEC devices, the main working principles of these devices must be understood. The main theory regarding PEC devices includes the electrochemical reaction basics, the semiconductor physics involved and the interaction between the two. In addition, it is important to consider the nature and spectrum of the solar irradiation. Finally, an idea on how to define the performance of the devices based on these concepts needs to be explored. This section explores all these concepts, starting by the nature of the solar energy, followed by the semiconductor definition and properties, the chemical reactions, and the interactions between these two sections.

SOLAR SPECTRUM AND SOLAR ENERGY

Both solar cells and photoelectrodes for water splitting are based on the photovoltaic effect, generating electron-hole pairs due to absorption of visible or other electromagnetic radiation by a semiconductor material. The photovoltaic effect can be further defined as "the generation of a potential difference at the junction of two different materials in response to [electromagnetic] radiation." (Zeman, 2011) Electromagnetic radiation is described by photons, namely defined as packages of energy with a value of (Zeman, 2011)

$$E = \frac{hc}{q\lambda}$$

Equation 1

where E is the energy of the photon, h is Planck's constant, c is the speed of light at vacuum, q is the elementary charge and l is the wavelength of the incident light.

The Sun releases enormous amounts of electromagnetic radiation into the solar system in the form of photons. The total power density of the solar radiation at the Earth in a perpendicular plane to the Sun, outside the earth atmosphere is 1353 W/m2. However, once it reaches the earth surface, this radiation is attenuated (Zeman, 2011). The Air Mass (AM) is a measure of how absorption in the atmosphere affects the spectral characteristics of the solar radiation reaching the Earth's surface. The Air Mass number is given by (Gray, 2003)

$$Air\ Mass = \frac{1}{\grave{e}os}$$

Equation 2

where θ is the angle of incidence.

A widely used standard for comparing solar cell performance is the AM1.5 spectrum, normalized to a total power density of 1 kW/m2 and can be roughly approximated to the radiation spectrum of a black body at 5762 K (Figure **1**). The spectral content of sunlight at the Earth's surface also has a diffused (indirect) component, due to scattering and reflection in the atmosphere and surrounding landscape, and can account for up to 20% of the light incident on a solar cell. This diffused light can still produce power in the solar cell, but not as efficiently as direct solar radiation (Gray, 2003).

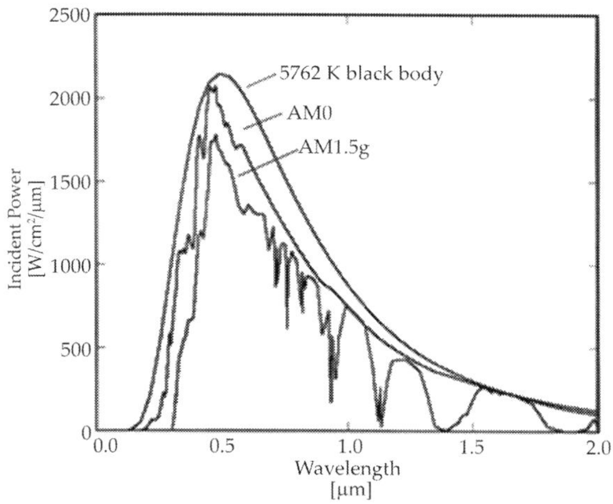

Figure 1. Radiation spectrum for a black body at 5762K, an AM0 and an AM1.5 global spectrum. From (Gray, 2003)

The study of the solar spectrum is especially important in photoelectrochemical (PEC) cells, since the main goal of these devices is to maximize its use. The photoelectrode material will absorb the most energetic part of the spectrum. The rest of the visible light, which has lower energy photons, will reach the solar cell. Therefore, the solar cell must be carefully designed in order to absorb the maximum part of the modified spectrum, especially in the lower energy region, and at the same time give enough voltage to the electrodes for the electrolysis to happen.

SEMICONDUCTOR CONSIDERATIONS

In PEC systems, semiconductors are used as electrolytes in order to create the needed potentials to drive the reactions described in the previous section. (van de Krol & Gratzel, 2012) The explanations of (Smets, et al., 2016) and (Balkanski & Wallis, 2000) of semiconductor physics fundamentals have been taken as a base for the explanations in this section. Semiconductors are materials whose electrical properties lie between those of insulators and conductors. This can be represented by the energy properties inside the atoms, as shown in **Figure 2**. The outermost electrons in the atoms can have energies either in the valence band, where they are strongly tied to the nucleus, or the conductive band, where they are relatively free. In between these energy levels, there is a series of forbidden levels called band gap energy. (Hybertsen & Louie, 1986) Metals, semiconductors and insulators can be differentiated by the band gap energies as further illustrated in **Figure 2**.

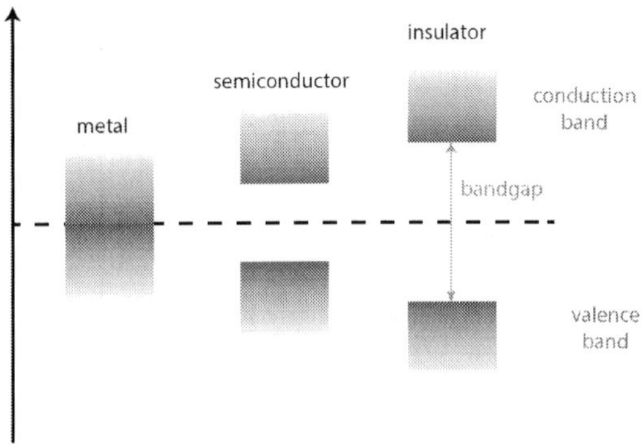

Figure 2. Schematic representation of the energy levels of metal, semiconductor and insulator materials.

Thus, in the case of metals, very little energy if any is needed to detach an atom from the forces of the nucleus into the conduction band, explaining the high conductivities and low resistivity of these materials. On the other hand, the energy needed to liberate an electron from the bonds with the nucleus in an insulator is considered so large that it effectively does not allow for any flow of electrons, and thus the high resistivity. In semiconductors, the bandgap energies lie in the middle, where it is possible to break the bond between outer electrons and atom nuclei, but some energy is needed for this to happen. (Balkanski & Wallis, 2000)

In this section, the main properties of semiconductors, including the bandgap energy, the possible doping profiles, the transport properties and the possible different junctions that can be formed between different materials.

Bandgap Energy

The outermost electrons in an atom are called valence electrons. They are involved in covalent bonds between atoms, and they have allowed energies bond to the nucleus defined as the valence band (VB). When an electron absorbs a certain amount of energy, it can be excited from the covalent bonds to the conduction band (CB). In between these two bands, there is a band of forbidden energy levels, where the electrons are not in a stable energy level and would try to either gain energy to reach the conduction band or release energy to reach the valence band. The maximum possible energy of the valence band is denoted as EV, and the minimum energy level of conduction band is called EC. The difference between these two energy levels is called the band gap energy, E_g, or simply the bandgap. Mathematically, it can be expressed as

$$E_g = E_C - E_V \quad \text{Equation 3}$$

Another important parameter is the Fermi energy, EFi. It is defined as the electrochemical potential of the free electrons in a material. Intuitively, it can also be pictured as the averaged energy of the electrons inside the material. The Fermi energy level of an intrinsic material is close to the midgap, and only a slight shift is caused by the difference of effective densities between the conduction band and the valence band. Mathematically, the Fermi energy level can be derived from the Fermi-Dirac distribution at a certain temperature, which is expressed as

$$E_{Fi} = \frac{E_C + E_V}{2} + \frac{kT}{2} ln\left(\frac{N_V}{N_C}\right) \quad \text{Equation 4}$$

where k is the Boltzmann's constant (k = 1.38 10-23 J/K), T is the temperature, NV is the effective density of charge carriers in the valence band, and NC is the effective density of charge carriers in the conduction band.

The Fermi-Dirac distribution results are represented in **Figure 3**, where the density of states (DOS), the Fermi-Dirac distribution and the electron and hole densities are represented. Regarding the DOS, it can be defined as the states in the conduction and valence bands that can be occupied by charge carriers. Ideally, there would be no states in the band gap that can be occupied by charge carriers, as shown in **Figure 3**. The Fermi-Dirac distribution, in the middle in **Figure 3**, represents a mathematical tool to describe the probability of a charge carrier to have certain energy. By combining both, the density of electrons and holes in both the valence and conduction band can be obtained, as illustrated in the right side of **Figure 3**. From this figure, it can be seen that the charge carriers tend to accumulate on the band edges, since these states are the ones that need the least energy. Finally, in all these figures the Fermi energy is represented in the middle of the band gap.

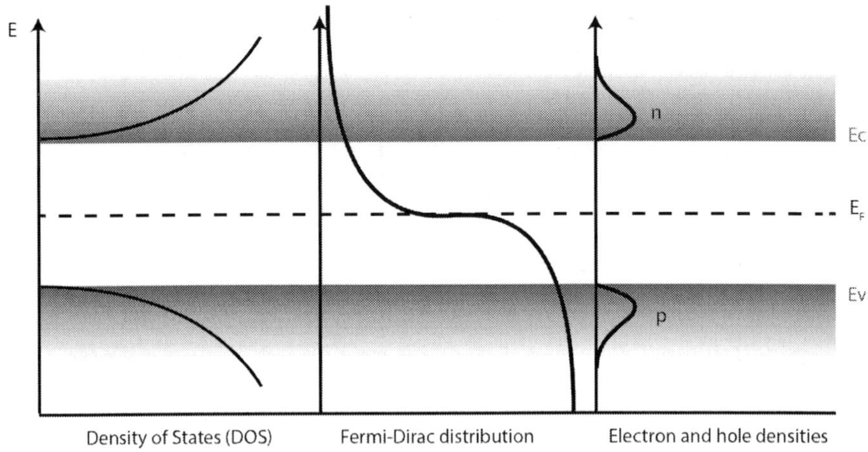

Figure 3: The representation of the density of states (DOS) function in conduction and valence band (left), the Fermi-Dirac distribution (center) and the electron and hole densities (right). Adapted from (Smets, et al., 2016)

Another parameter expressed by the Fermi-Dirac distribution is the fact that temperature has an effect on these energy levels and charge carrier densities. It must be noted that, when temperature is increased, more charge carriers will appear as a result of the energy provided to the material. Thus, at a temperature of 0 K, all electrons are below the Fermi level. As the temperature is increased, some electrons move to levels higher than the Fermi energy level. Ultimately, this movement will make charge carriers to move to the conduction band and the valence band, defining the band gap.

The bandgap of the semiconductor is also a key factor to determine the wavelength of the light absorbed in the material. In the case of PEC, an efficiency theoretical limit based on how much light can theoretically be absorbed by the particular bandgap can be defined. This is shown in **Figure 4**, where also the thermodynamic limit of 1.23 V for the water splitting reaction is pointed out. This means that, thermodynamically, the semiconductor would need to have a bandgap above 1.23 eV to produce the necessary voltage, while being as low as possible to absorb the maximum amount of light. (Chen, et al., 2013)

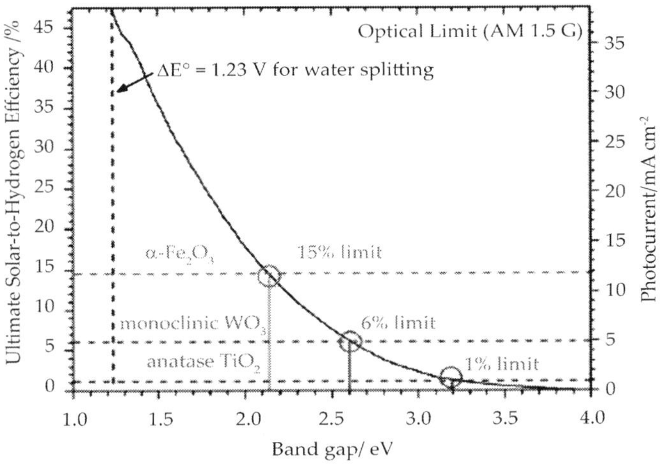

Figure 4: Theoretical maximum solar-to-hydrogen efficiency based on the semiconductor material bandgap. Note that this graph only considers the thermodynamic principles, and it is equivalent to the Schockley-Queisser limit for solar energy conversion. From (Chen, et al., 2013)

There are several ways to use these basic properties to come close to the efficiency limits, improving the chemical and electrical properties of such systems. These strategies are explored further in this section.

Doping

The concentration of charge carriers in the semiconductor can be tuned by doping the material. Doping means that different atoms are introduced in the structure. For example, in the case of crystalline silicon, it is common to introduce boron atoms as acceptors and phosphorous atoms as donors. These atoms would substitute silicon atoms in the crystalline structure. They are usually chosen because they have a similar size to the silicon atom, but their valence electrons change from four (Si) to either three (B) or five (P). When an atom of boron is introduced in the lattice, it will form three bonds with the surrounding silicon atoms. However, the fourth silicon does not have another valence electron to bond with, and would try to capture one of the free electrons in the lattice. On the other hand, when a phosphorous atom is introduced into the lattice, it will form four bonds with the four surrounding silicon atoms. However, phosphorous has a fifth valence electron that is not able to bond with any of the surrounding atoms. Thus, this electron will become weakly bonded to the phosphorous nucleus, and would easily move to the conduction band, becoming one extra free electron.

When looking at the energy levels previously described, the Fermi levels would shift, since the concentration of free electrons in the conductive band has changed. Thus, for materials with donor atoms as dopants, the Fermi level will be closer to the conduction band, while in the case of having acceptor atoms as dopants, the Fermi energy level will be closer to the valence band. The material containing donor atoms would also be called n-type, due to the increased concentration of negative charge carriers. The material with acceptor atoms would be called p-type material. The band gap diagrams of these two material types are illustrated in **Figure 5**, including the difference in the Fermi energy levels and the different charge carrier concentrations. It is also important to notice that, even though additional charge carriers have been created, the material still maintains charge neutrality.

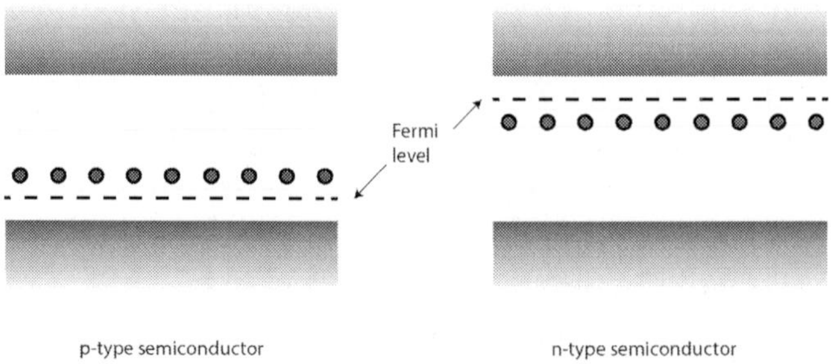

Figure 5: Band diagram of p-type (left) and n-type (right) semiconductors. Note that the Fermi levels are not in the middle of the band gap due to the extra holes and electrons from the doping atoms, respectively.

It is important to mention that this effect of shifting the Fermi level away from the middle of the mid-gap can be achieved in other ways. In particular, there are some semiconductor materials, especially oxides, that are naturally n or p type. That would be due to the different connections and bonds between the atoms of the compound, which can create either acceptor or donor spots, creating the illusion of material doping. (Alexander, et al., 2008)

Transport Properties

The electrical currents are generated due to the transport of electrons and hole inside the material. The transport mechanisms that these charge carriers can use are drift and diffusion. (Smets, et al., 2016)

Drift. In this mechanism, charge carriers move due to an electric field, since they are charged particles. The holes (or positive charges) would move in the direction of the electric field, while electrons would move in the direction opposite to the electric field. The effects of an electric field on the band structure of the semiconductor are reflected on **Figure 6**. In this image, the electrons will tend to go in the opposite direction of the electric field, while holes would go in the same direction.

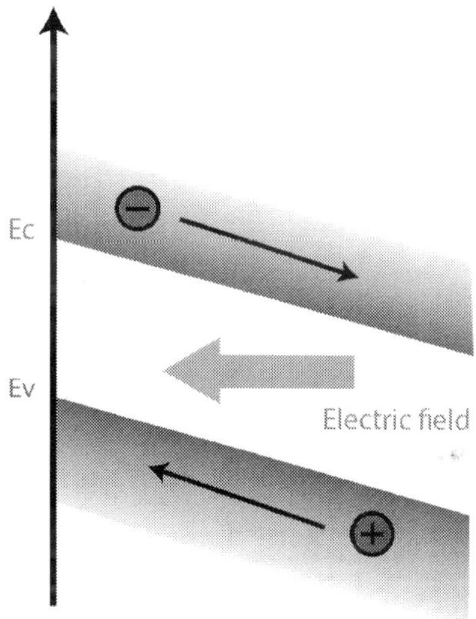

Figure 6: Band diagram of a semiconductor with an electric field

This mechanism can be disturbed due to vibrations in the lattice atoms or ionized impurity atoms inside the material lattice. Thus, it is important for devices depending on drift mechanisms to have high quality materials with as little impurities as possible, which will be partly determined by the fabrication technique of the semiconductor material.

The drift mechanisms can be characterized by the velocities of charge carriers and resulting electric currents. The drift velocities of these charge carriers are directly proportional to the electric field by a proportionality factor, called mobility.

The final current produced by drift can be defined by the movement of both electrons and holes as

$$J_{drift} = e\left(n\mu_n + p\mu_p\right)\bar{E}$$

Equation 5

where e is the elemental charge, and n and p are the amount of electrons and holes in the material, respectively, μ is the proportionality factor called mobility, and is the electric field.

Thus, one of the main parameters to improve if this mechanism is used is the electric field inside the semiconductor material. That can be done by including doped layers within the device structure, or by applying a bias voltage. Thus, the semiconductor device design is important to achieve good performances when using drift mechanisms to control the electrical current.

The mobility is another of the most important characteristics describing the electrical properties of a semiconductor. It can be understood as the easiness of charge particles to move through a semiconductor. This characteristic would depend, among other parameters, on the quality of the material, the impurity concentration, the doping concentrations and the temperatures. In addition, it is important to note that not all charge carriers have the same mobility in a certain material, especially if it is doped. Thus, if the main transport mechanism is drift, all these parameters must be taken into account.

Diffusion. In this mechanism, the charge carriers move due to particle concentration gradients. Thus, this mechanism does not depend on the charge of electrons and holes but in their mass and density inside the material. In general, particles tend to spread from regions of high concentration to regions of low concentration as a result of random thermal motion. This concept is illustrated in **Figure 7**.

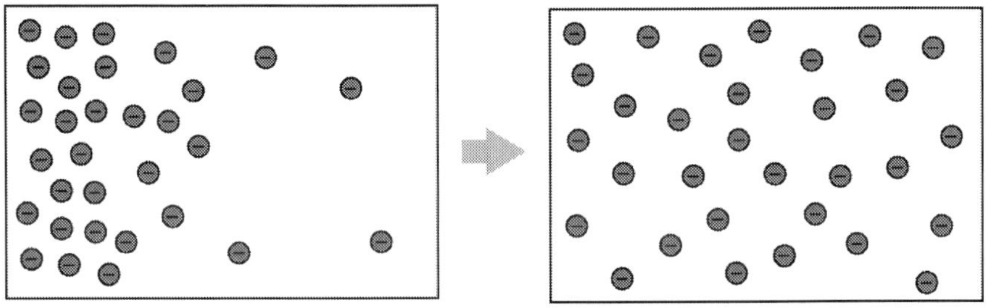

Figure 7. Illustration of the diffusion process of electrons inside a material with time

The diffusion velocities are proportional to the concentration gradients, where the proportionality constants are called diffusion coefficients. Mathematically, it can be expressed as

$$J_{n,diff} = eD_n \nabla_n$$ Equation 6

$$J_{p,diff} = -eD_p \nabla_p$$ Equation 7

where $J_{i,\ diff}$ are the current densities for electrons and holes, e is the elemental charge, Dn and Dp are the diffusion coefficients for electrons and holes, respectively, and and are the concentration gradients of electrons and holes, respectively.

The total diffusion current will be the sum of the two currents

$$J_{diff} = e\left(D_n \nabla_n - D_p \nabla_p\right)$$ Equation 8

Total electrical current. Finally, the diffusion current can be combined to the drift current to obtain the final overall current that can be achieved in the semiconductor.

$$J = J_{drift} + J_{diff} = e\left(n\mu_n + p\mu_p\right)\bar{E} + e\left(D_n \nabla_n - D_p \nabla_p\right)$$ Equation 9

From this equation, it is clear that the current produced in a semiconductor is dependent on the concentration of charge carriers in the semiconductor, the mobility of the carriers and the diffusion coefficients of the charge carriers inside the material. Nevertheless, it is likely that one of the two mechanisms is predominant over the other, giving more importance to either the mobility or the diffusion coefficient.

Semiconductor Junctions

In order to create a flow of charge carriers in semiconductors, it has been discussed that there must be either a drift created by and electric field or diffusion between regions with different carrier concentrations. Thus, there is a need to create structures and device architectures that enhance one of these mechanisms. The semiconductor/electrolyte interface already creates a junction, which would be further explained in further sections. However, semiconductor/semiconductor junctions can also be created for better charge separation and collection. To achieve this, different materials of different doping can be used to either create electric fields or carrier concentration profiles. The most important ones are the p-n junction and the metal-semiconductor junctions, which will be discussed here.

p-n junctions. They can be made out of the same semiconductor material, such as crystalline silicon, called homojunctions, or of two chemically different semiconductors, called heterojunctions. By creating a p-n junction, an electric field is formed. Charge neutrality would be conserved, where the large concentration of free electrons in the n-type is compensated by the holes in the p-type semiconductor. This created a very large difference in electron concentration, creating a diffusion current of electrons from the n-type material into the p-type material. Next to the interface, the region is poor of charge carriers due to that diffusion current, creating the so-called depletion region. This concept is illustrated in Figure 8. Note that the Fermi energy level remains constant throughout the whole structure.

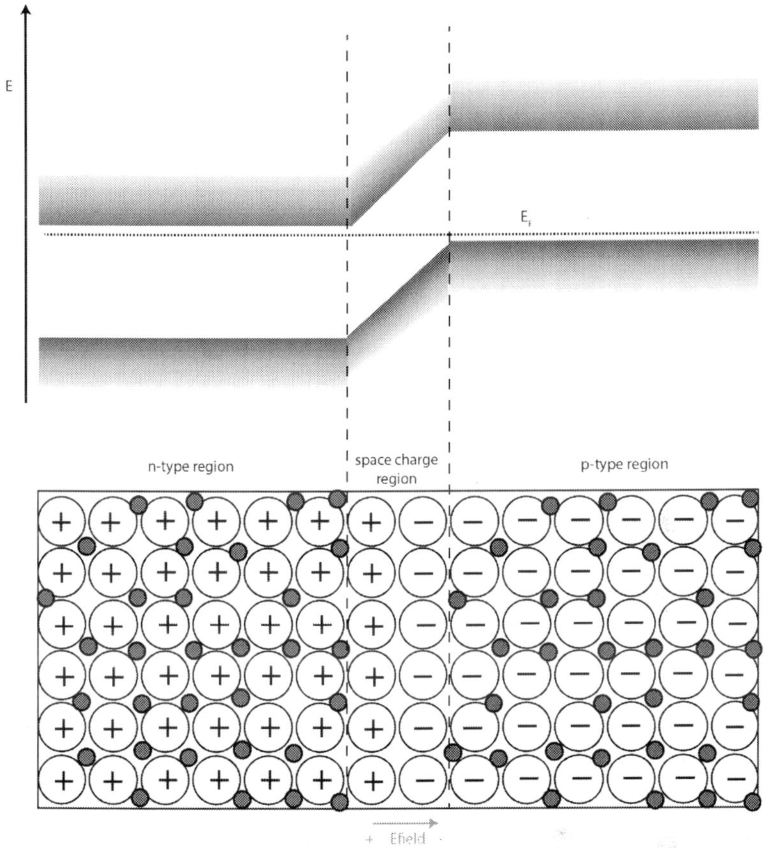

Figure 8: Band diagram and schematic indicating the formation of a space-charge region when an n-type and a p-type semiconductor are brought together and forma a junction.

When an external voltage is applied in a p-n junction, the electrical field across the junction will change, taking it out of its equilibrium, as shown in Figure 9. When the voltage is negative with respect to the built-in voltage, the potential difference will increase, widening the space-charge region. On the other hand, if the applied voltage is positive, the potential difference will be reduced and the space-charge region would become narrower This positive potential, also called forward bias, will lead to higher concentration of minority carriers at the edges, which is called minority-carrier injection. The carrier concentration gradient causes diffusion.

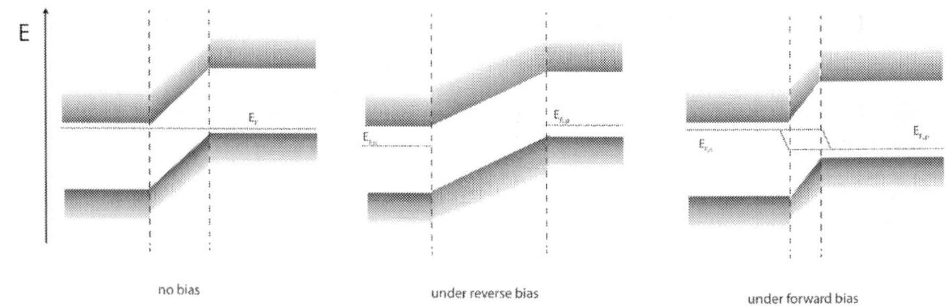

Figure 9: Energy band diagram of the space-charge region under no bias (left), reverse bias (center) and forward bias (right). Adapted from (Smets, et al., 2016)

If an intrinsic layer is introduced in between, creating a p-i-n junction, this would act as a capacitor and the electric field would be extended. Then, drift from the electric field would be the main mechanism. Under equilibrium, the Fermi levels will be constant throughout the junction, and the conduction and valence bands would maintain their continuity, creating a band bending across the depletion region.

Heterojunctions. These are junctions between different materials, typically related to two different semiconductors. In this case, the electron affinity and bandgap are not the same on both sides of the junction, and thus the potential barriers that can be formed must be taken into account. However, the Fermi energy level remains constant across the junction, the same as in the homojunctions described before. This is displayed in Figure 10. In addition, the vacuum energy should be continuous, but the conduction and valence bands are not necessarily continuous. The built-in voltage of the heterojunction would then be the difference in between the two work functions.

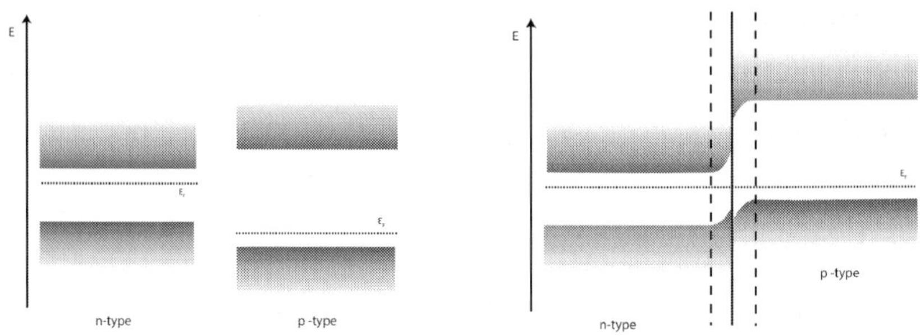

Figure 10: Band diagram of the two materials with different bandgaps and affinity (left), and the heterojunction formed (right). Adapted from (Smets, et al., 2016)

Metal-semiconductor (MS). These junctions are very important in electronic devices to connect the different elements inside the devices with the external loads by metal cables. Depending on both the metal and the semiconductor characteristics can lead to two types of junctions: the rectifying type, and the non-rectifying type (ohmic contact). Among the metal characteristics that determine this is the work function, while the semiconductor is characterized by the electron affinity. For an ideal contact, there are two main requirements: the constant thermal equilibrium of the Fermi energy throughout the junction, and a continuous vacuum level. The band diagram of this junction is displayed in Figure 11.

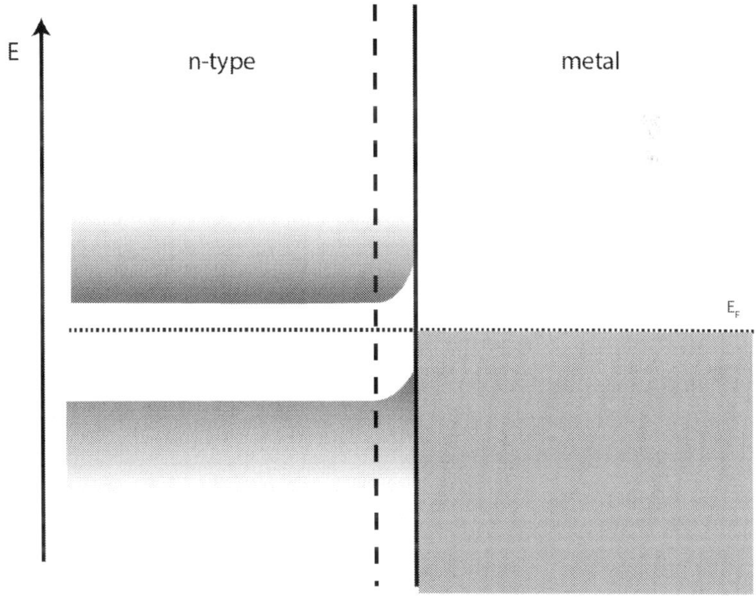

Figure 11: Band diagram of a metal-semiconductor junction for an n-type semiconductor. Adapted from (Smets, et al., 2016)

In these junctions, a Schottky barrier is often formed. Unlike the previous junctions described, here the majority carriers are the responsible for the current transport. Moreover, there would be an energy barrier for this transport, and no current would be transported without a voltage applied. If there is a forward bias, the energy barrier decreases and the density of charge carriers increases. On the other hand, if the bias is reversed, the current would flow from the metal to the semiconductor.

Recombination of Electron-hole Pairs

The recombination of electron-hole pairs is usually an undesired effect that happens inside the semiconductor, reducing the amount of current in the material. Depending on the semiconductor properties, there would be different recombination mechanisms. The most important ones would be discussed here.

Direct thermal recombination. This often happens in semiconductors such as gallium arsenide. It is caused by the intrinsic vibrational states inside the materials, which can lead to both generation and recombination of charge carriers. In this case, the electrons simply decay to the valence band, liberating the excess energy as heat. This is illustrated in the leftmost picture in Figure 12. In this case, it can be consider that the direct recombination rate is proportional to the concentration of electrons in the conduction band and to the concentration of holes in the valence band. However, this recombination mechanism is not very significant in most semiconductor devices.

Shockley-Read-Hall (SRH) recombination. In this case, the recombination does not directly happen from bandgap to bandgap, but it is facilitated by a mid-bandgap impurity or lattice defect. These impurities or defects can also be used as recombination centers, and they introduce energy levels in the forbidden gap, called trap states. The electron would be trapped in this defect, and the hole would be attracted by the electron, causing the recombination and releasing heat. This is illustrated in the center of Figure 12. This recombination mechanism must be highly taken into account, since they are the most important mechanism in most semiconductor devices and operational states.

Auger recombination. This recombination mechanism is found in direct semiconductors. It involves three particles instead of two (electrons and holes). In this case however, the hole and electron maintain its momentum and energy, and the extra energy is transferred to another electron or hole in the surrounding. If the energy is transferred to an electron, it would be excited to higher energy levels. When it relaxes again, the energy would be transferred to the vibrational energy or the lattice, and finally converted into heat. If the third particle is a hole, it would sink further into the valence band, and when it rises again would have a similar effect. An example of this mechanism when using two electrons and one hole is illustrated in the right picture of Figure 12. This type of recombination highly depends on the charge carrier densities, and thus in the doping levels, being more prevalent at high doping levels.

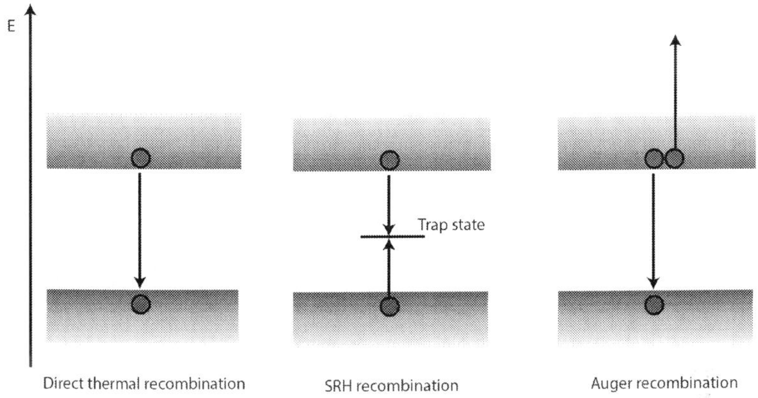

Figure 12: Band diagrams of the mechanisms for direct thermal recombination (left), Shockley-Read-Hall (SRH) recombination (center) and Auger recombination (right).

Surface Recombination. So far, all the mechanism studied affects the bulk of the material. However, there is also recombination on the surface. This is due to the atoms in the surface, where many valence electrons cannot find a partner and result in the so-called dangling bonds. These dangling bonds are considered defects that create trap states with in the bandgap, leading to SRH recombination. If the bulk properties of the semiconductor are of good quality, the recombination of the device might be dominated by surface recombination, and to improve the performance this surface needs to be passivated, meaning that the dangling bonds must be removed by an additional treatment or by depositing an additional layer of a suitable material. Another way to avoid surface recombination is to create a highly doped region underneath the surface, creating a barrier for minority charge carriers

Chemical Reactions

So far, the main factors about light properties, light absorption and other characteristics of the semiconductor have been reviewed. However, it is also important to look at the chemical reaction happening at the electrolyte. Photoelectrochemical water splitting is based on the electrolysis of water, where water is split into hydrogen and oxygen by the application of a flow of electrons. The reaction can thus be defined as a reduction-oxidation reaction, or redox reaction. These reactions are characterized by the reduction (gain of electrons) of one species and the oxidation (loss of electrons) of another. Since these two reactions can occur in different electrodes, they are often called half-reactions. Each half reaction is associated with an electrical potential that the electrons would need to cover in order for the reaction to happen. In this case, the two half reactions can be defined as (Chen, et al., 2013)

Cathode $H_2O + 2h+ \rightarrow 2H^+ + \frac{1}{2} O_2$ $E^0_{ox} = 1.23$ V

Anode $2H^+ + 2e- \rightarrow H_2$ $E^0_{red} = 0$ V

Overall $H_2O \rightarrow H_2 + \frac{1}{2} O_2$ $E^0 = 1.23$ V

Equation 10

Note that the 1.23 V here discussed represents the minimum thermodynamic voltage for splitting water. However, other losses can appear in the system, such as the charge transport and injection into the electrolyte, kinetics of the half reactions (OPHER, OPOER), etc. These overpotentials can be minimized by using the right catalyst, semiconductor and other strategies. (Chen, et al., 2013)

However, it also must be taken into account that these reaction mechanisms also depends on conditions of the medium such as the electrolyte pH. The previous equations happen when the conditions are acid, with a low pH, as the abundance of H+ shows. If the medium is basic, with high pH, the reactions will change as

Cathode $2H_2O + 2e- \rightarrow 2OH- + H_2$ $E^0_{ox} = -0.83$ V

Anode $2OH^- + 2h+ \rightarrow \frac{1}{2} O_2 + H_2O$ $E^0_{red} = -0.40$ V

Overall $H_2O \rightarrow H_2 + \frac{1}{2} O_2$ $E^0 = 1.23$ V

Equation 11

In order to better understand each of these processes involved in the electrochemical reaction, each element will be studied in detail in the following subsections.

Potentials in the Electrochemical Cell

Each electrochemical potential is associated with a chemical potential related to the thermodynamics of the solutions and species involved. The basic definition of a chemical potential is given as (Memming, 1998)

$$\mu_i^{\alpha} = \left(\frac{\partial G}{\partial n_i}\right)_{T,p,n_{j\neq i}}$$

Equation 12

where μ_i^{α} is the standard chemical potential for the i species in phase α, G is the Gibbs free energy and ni is the number of moles of that species in the particular phase being studied α. This definition is valid if the temperature T, the pressure p and the concentration of the species n are kept constant. In the ideal solution, this can be translated to (Memming, 1998)

$$\mu_i = \mu_i^0 + RT\ln\left(\frac{c_i}{c_i^0}\right)$$

Equation 13

where μ_i^0 is the standard chemical potential for $c_i = c_i^0$. However, for real solutions this would not be accurate enough, since it only takes into account interactions between the ions and the solution, and neglects ion-ion interactions. Thus, a correction can be introduced by defining the activity coefficients f, as (Memming, 1998)

$$\mu_i = \mu_i^0 + RT\ln\left(f\frac{c_i}{c_i^0}\right) = \mu_i^0 + RT\ln\left(\frac{a_i}{a_i^0}\right)$$

Equation 14

where a_i is the activity of the species.

Finally, if the species are charged, the electrical energy required to bring a charge to the site of the species must also be taken into account. Thus, the electrochemical potential can be defined as (Memming, 1998)

$$\phi_i = \phi^0 + RT\ln\left(\frac{j_{c,i}}{j_{c,i}^0}\right)$$

Equation 15

where j_c is the current density. This can be related to the chemical potential as (Memming, 1998)

$$\overline{\mu_i} = \mu_i + z_i F\phi$$

Equation 16

where $z_i F \varphi$ is the potential energy of the charged species, with F as the Faraday constant, z_i the charge number of the species and φ the potential in volts. It is important to notice that, at equilibrium, the electrochemical potential would be maintained constant over all the phases as long as the substance is exchangeable between phases. Thus, in that case the rate of the reaction must be equal in both directions, resulting in (Memming, 1998)

$$\sum_i \upsilon_i \bar{\mu}_i = 0 = \Delta G$$

Equation 17

where ν_i of the product must be positive and those of the reactants are negative.

These differences in potential result in a potential profile at the interface between the electrode and the electrolyte, as illustrated in Figure 13. Note that impurities at this interface can vary the profile, changing the interactions between electrode and electrolyte, and affecting the reaction rates. Thus, this interface it is very important and will play a crucial role in the photoelectrochemical reactions that would be later studied in more detail. (Bard, et al., 1993) It must also be noticed that these potential differences cannot be measured accurately, since for that another electrode would have to be introduced, affecting the measurement itself. Thus, although this is a very important characteristic for the overall system, it is also very difficult to characterize accurately.

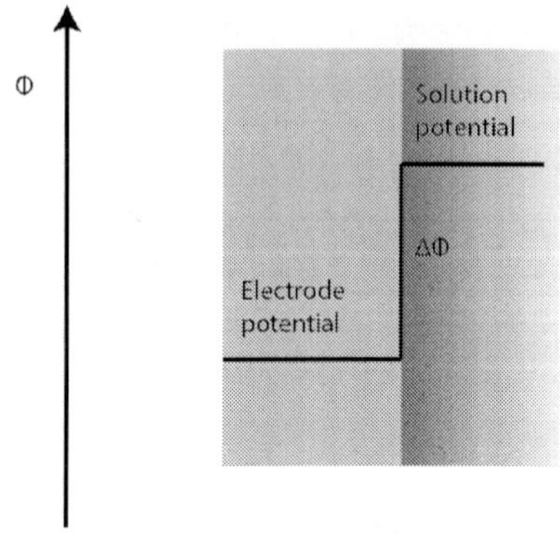

Figure 13: Potential profile across a metal-liquid interface. Adapted from (Memming, 1998)

When these electrodes are applied into an actual device, a certain potential difference is thus produced between the electrode and the electrolyte. This potential difference, also called cell voltage, is one of the main parameters used to define and improve the system performance, so it is very important to determine which the most important parameters that can affect it are. In addition, as can be seen previously, the potentials with a 0 superscript are considered as reference potential. However, to better understand these concept, they will be further explored.

Cell voltages. The electrochemical cells can also be defined by the voltages that appear between the two electrodes, which are directly related to the chemical and electrochemical potentials of the reactions. These voltages depend on the half-reactions occurring at each interface, as already mentioned. If a membrane is introduced in between the two electrodes, an equilibrium cannot be reached, since the exchange of ions between the two partial systems is not possible. However, equilibriums are then established in each of the half cells between the electrodes and the ions in the solution, creating different potentials in each of the electrodes and thus an electrical potential appears between the two electrodes. This potential can be defined as (Memming, 1998)

$$U = \Delta\overline{\mu_e}F = -\frac{\Delta G}{F}$$ Equation 18

These potentials are also illustrated in Figure 14. As seen in this figure, the cell voltage is defined as the potential difference between the two electrodes, without the electrolyte potential playing a major role. However, that can be misleading, since in reality the solution potential would affect the charge injection and transport within that layer, as it will be further explained in the following sections.

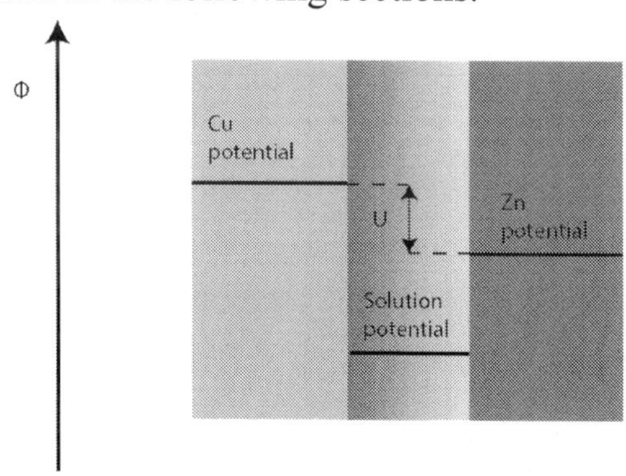

Figure 14. Potential profile example for a Cu-electrolyte-Zn system. Adapted from (Memming, 1998)

Reference potentials. In order to define these different potentials, a point of reference must be taken. The most common reference electrode taken is the standard hydrogen electrode (SHE) or normal hydrogen electrode (NHE). (Neuman & Thomas-Alyea, 2004) The reaction considered to have a potential of 0 V would then be (Memming, 1998)

$$2H^{+} + 2e^{-} \rightarrow H_2 \qquad \text{Equation 19}$$

and the electrochemical potential would be given as

$$\overline{\mu_e} = \frac{1}{2}\mu^0(H_2) - \mu^0(H^+) - RT\ln\left(\frac{a(H^+)}{p^{1/2}(H_2)}\right) - F\phi^{sol}$$

Equation 20

The system, including a Pt counter electrode, is depicted in Figure 15, where H_2 is flushed to the system in one of the electrodes. The hydrogen reaction potential is measured at the Pt electrode and compared to the voltage at the working electrode. Thus, the potential difference reported against NHE assumes this hydrogen reaction happening at one side.

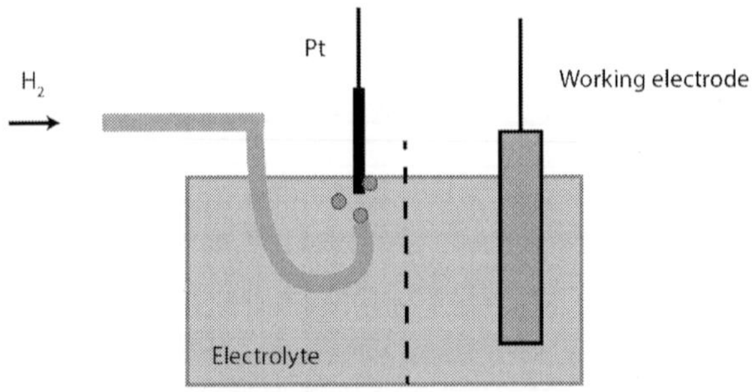

Figure 15. Hydrogen reference electrode system (normal hydrogen electrode, NHE). Adapted from (Memming, 1998)

Since hydrogen is normally a gas, the concentrations can be defined by the partial pressures of the gases p. The potential of the half-cell can then also be expressed as

$$U_{ref}(H_2 / H^+) = -\overline{\mu_e}F \qquad \text{Equation 21}$$

At standard conditions, p1/2(H_2) = A(h^+) = 1, resulting in a reference voltage defined as

$$U^0_{ref} = -F^{-1}\left[\frac{1}{2}\mu^0(H_2) - \mu^0(H^+)\right]$$

Equation 22

These are the conditions accepted as a reference point, the so-called standard hydrogen electrode (SHE). Thus, if another electrode is included, the overall voltage can be defined as

$$U_E = U_{syst} - U^0_{ref} = \frac{\Delta\mu}{F} = \frac{\Delta G}{zF}$$ Equation 23

The overall system regarding the energy bands is illustrated in Figure 16. It must be noted that these potentials depend on the concentration of protons, i.e. the pH of the solution. The reference voltages are defined at pH_0, and they would change 60 mV for each pH unit variation.

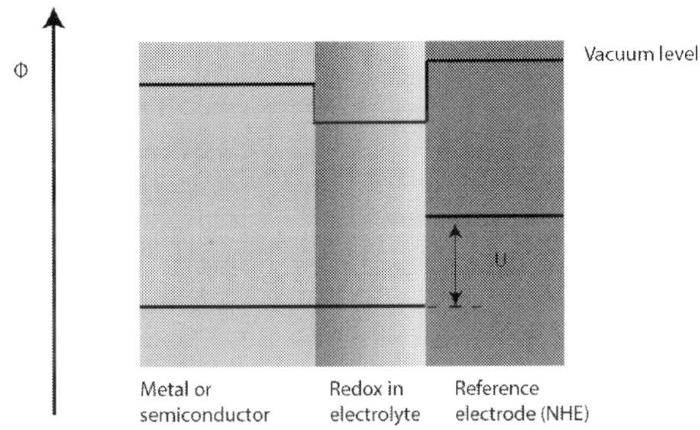

Figure 16. Energy diagram for a metal or semiconductor redox system-reference electrode system. Adapted from (Memming, 1998)

In these systems, the Fermi energy levels of the reactions, represented by the bottom solid lines, and characterizing the energy of the charge carriers within the materials can be defined based on the concentrations of the different species, as

$$E_{F,\, redox} = E^0_{F,\, redox} - kT \ln\left(\frac{c_{ox}}{c_{red}}\right)$$ Equation 24

where the system is considered to be in equilibrium when the Fermi energy level of the semiconductor is equal to the one for the redox reaction. These energy levels are also illustrated in Figure 17, where this schematic leads to a certain potential difference across the interface, which will be discussed in detail in subsequent sections.

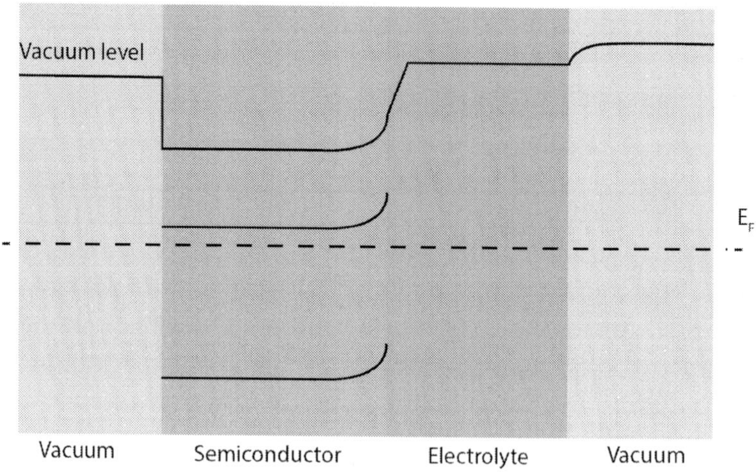

Figure 17: Energy diagram for the semiconductor-vacuum, semiconductor-liquid and liquid-vacuum interfaces. Adapted from (Memming, 1998)

Based on these definitions, the redox potential can also be related to the concentration of the different species by the famous Nernst equation

$$U_{redox} = -\frac{E^0_{redox}}{e} = U^0_{redox} + kT\ln\left(\frac{c_{ox}}{c_{red}}\right)$$ Equation 25

where U^0_{redox} is the standard redox potential at $c_{ox} = c_{red}$.

In addition, although the SHE is used as a reference for calculations, it is not a practical system to measure potentials experimentally. Thus, in those cases other electrodes such as the Ag/AgCl electrode is used, with a potential of 0.199 v vs. NHE and a very stable nature. (Memming, 1998) However, caution should be exercised when comparing redox results with different reference electrodes, and thus it is advisable to use the same reference electrode for all measurement, and possibly being the same as the standard of the field. (Pavlishchuk & Addison, 2000)

Electrolyte

The electrical current in an electrolyte is carried by ions, unlike in the case of metals and semiconductors. These ions appear from added species such as KCl or Na_2SO_4, which would dissociate in solution to form these ions, both positive and negative. Unlike in semiconductors, where electrons and holes form, but one of them is dominant, these ions are equally distributed throughout the solution. The conductivity of the solution describes the easiness of ions to travel within an electric field in the solution, and can then be given by (Memming, 1998)

$$\sigma = F\sum_i |z_i| \mu_i c_i$$

Equation 26

Where z_i is the charge of each ion specie, μi is the mobility of the ion, ci is the concentration of ions, and F is the Faraday constant 96500 C mol-1. Based on these properties the desired amount of ions to be introduced in the solution in order to assure conductivity in the electrolyte can be calculated. In addition, it must be taken into account that these species can also play a role in the reaction, as reactants, how they interact with the acids or bases that might be present. Thus, species chemically inert for the reaction at hand must be chosen.

In addition, ions and molecules can also move through the electrolyte by diffusion, which is an important mechanism around the electrode, where the different molecules and ions need to move in order for the reactants to reach the surface and the products to move away from the surface of the semiconductor. This movement occurs due to a difference in concentrations between the interface and the electrolyte, which facilitates these movements. The diffusion process can be described by Fick's law, as

$$D = \frac{kT}{e}\mu_i$$

Equation 27

where D is the diffusion coefficient, k is the Boltzmann constant, T is the temperature and e is the charge of an electron.

Electrodes

Entropic losses in the semiconductor must also be taken into account. These would be associated with photoemission, reflective and transmittance losses and non-radiative recombination, which are typical losses found in optical semiconductor devices and would be described more in detail in the following section. (Coridan, et al., 2015) In addition, the band alignment of the semiconductor and the electrolyte are crucial, and would also be discussed in further sections. (Chen, et al., 2013) Note that some of these properties apply only to the photoelectrode, and other properties are also relevant for the counter electrode, which is often based in metal materials. In summary, all these phenomena would add to the practical operational potential needed, which would be in the range of 1.6-2 V.

In particular, the surface of the electrode, both in band alignment and crystal orientation, heavily affects the performance of the overall system, mostly due to the kinetics between the electrode surface and the electrolyte and its reaction. (Lewerenz, et al., 1984) Moreover, it has been determined that the intermediate species and mechanism of the reaction might change depending on the electrode properties. (Park, et al., 2013)

Thus, based on this analysis, it is important not only to look separately at the semiconductor properties and the electrolyte and chemical reaction characteristics, but also at the interaction of the combination of both. (Linsbergler, et al., 1995) This will be tackled in the following sections, looking at the bare semiconductor and at the effect of including a catalyst on the surface. In addition, not only the performance of the electrode will be affected by this interaction, but also the stability of the material, and thus this should also be explored.

SEMICONDUCTOR/ELECTROLYTE INTERACTIONS

So far, the properties of both the solution and the semiconductors that compose PEC devices have been discussed. However, the most important feature is arguably the interaction between these two elements. Thus, this section focuses on the interactions occurring at the interface between the semiconductor material and the electrolyte solution.

There are several factors involved in the complete process of water splitting, illustrated in Figure 18. Mainly, the photon irradiation, electron-hole pair formation and charge transport are mostly related to the semiconductor material properties. The energy requirements associated with the minimum thermodynamic energy to split water are mostly related to the reaction itself and the electrolyte properties. However, the catalytic overpotentials for each of the half reaction, and the interfacial reactions and mechanisms are dependent on this interface properties. (Chen, et al., 2013) Effective methods to describe and characterize these properties must be explored and analyzed.

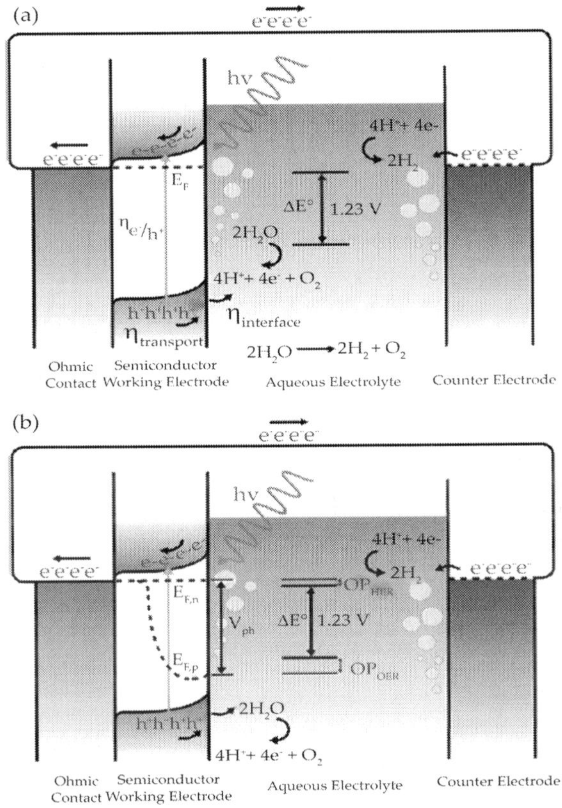

Figure 18: Band structure of an n-type photoanode water splitting device, illustrating a) the photon irradiation, electron-hole pair formation, charge transport, and interfacial reactions, and b) the energy requirements associated with the minimum thermodynamic energy to split water, catalytic overpotentials for each of the half reaction and photovoltage. From (Chen, et al., 2013)

One of the elements that is used to characterize this interface is the flat-band potential Efb of the photoelectrode. The sign of this value would mostly depend on the kind of electrodes used, and thus the reaction that evolves on its surface. For example, n-type semiconductors are used as photoanodes, to conduct the oxygen evolution reaction (OER). On the other hand, p-type semiconductors are used as photocathodes, to drive the hydrogen evolution reaction (HER), due to the difference in band alignment at the surface. Depending on if a semiconductor is n-type or p-type, the energy bands would shift in a different direction during open circuit condition upon illumination, as illustrated in Figure 19. The Fermi level of the bulk semiconductor would shift to more anodic potentials during illumination for p-type semiconductors and to more cathodic ones for n-type semiconductors. (Chen, et al., 2013) The flat band potential is related to these energy levels, and it is defined as the potential at which the Fermi energy levels are equal to the redox potentials, and thus there is no band bending. (Gratzel, 2001) Thus, it would have a different value for each of these materials, since the band bending goes in opposite directions in each of the cases explained.

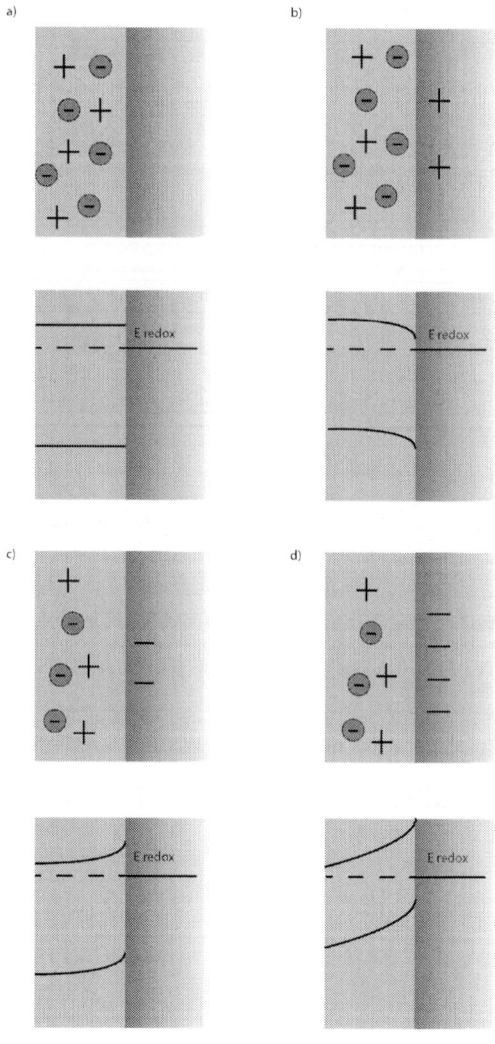

Figure 19. Schematic showing the electronic energy levels at the interface between an n-type semiconductor and an electrolyte containing a redox couple.

The four cases indicated are: a) flat band potential, where no space-charge layer exists in the semiconductor; b) accumulation layer, where excess electrons have been injected into the solid producing a downward bending of the conduction and valence band towards the interface; c) depletion layer, where electrons have moved from the semiconductor to the electrolyte, producing an upward bending of the bands; and d) inversion layer where the electrons have been depleted below their intrinsic level, enhancing the upward band bending and rendering the semiconductor p-type at the surface. From (Abdullah-Al-Mammun, 2010)

This affects the electron transfer process at the interface, which is often not perfect and thus creates losses due to kinetic overpotentials needed to drive the half-reactions. This increases the voltage requirements, which thermodynamically were defined as 1.23 V, to around 1.6-2.4 eV per electron-hole pair generated. (Walter, et al., 2010) That is why, as will be discussed later, multijunction devices are often needed, since one single semiconductor cannot produce such high voltages. In addition, the semiconductor/electrolyte interface properties also affect these kinetics, and thus these voltages can be lowered by focusing on those interfaces.

Thus, it has been shown that this interface is crucial in PEC devices. Here, the main mechanisms defining the performance at such interface are defined. The main factors explored in this book are the adsorption of the different reactants and products at the interface, the band alignment, the possible charge carrier accumulation at the surface, and the catalytic effects on the surface.

Adsorption

When a metal or semiconductor is in contact with a liquid, the ions and molecules may be adsorbed at the surface, facilitating or hindering the electrochemical reactions. This would depend on the molecules and the solid material, and thus it is important to consider the semiconductor and ion properties regarding this issue. (Delahay & Trachtenberg, 1956)

The adsorption process can be described as a chemical process in which bonds between these molecules and the surface. Some examples of these bonds are shown in Figure 20. (Memming, 1998) These examples in particular show the cases in which different reactants are attached on the surface, such as OH ions or water molecules, enhancing the overall electrochemical reaction. However, in other cases, such as carbon structures, the products are attached to the surface, blocking any further reaction processes happening at that location, and thus hindering the overall reaction. (Panella, et al., 2005) Thus, the surfaces must be optimized to adsorb the reactant species and desorb the products once the reaction has taken place.

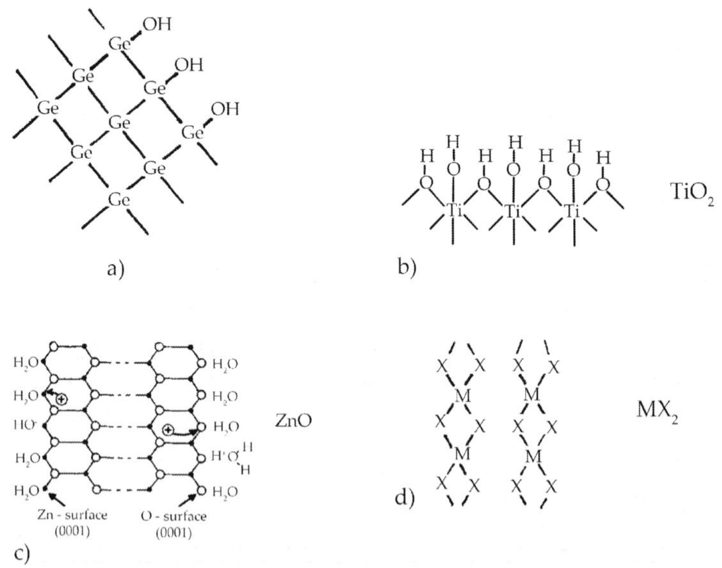

Figure 20. Surface structure of various semiconductors: a) Ge; b) TiO2; c) ZnO; and d) layered dichalcogenides (MX2). From (Memming, 1998)

It is also important to notice that the absorption process also depends on the orientation of the atoms at the surface, and thus, the crystal orientation of the metal or semiconductor would play an important role on the reaction rate. (Faguy, et al., 1990) The crystal orientation can be controlled by controlling the deposition conditions, tuning the parameters to obtain the desired structure. Moreover, nanopatterning can also aid in the adsorption process, (adachi, et al., 2004) making it an interesting technology to further implement and it will explored in following chapters.

Another way in which adsorption can happen is by bonding with an intermediate species rather than directly with the target species. For example, OH- ions tend to have difficulties when bonding with semiconductor surfaces. Thus, there would be F- ions that would bond with Ge and Si surfaces, which in exchange can bond further with the

OH^- ion. In addition to this chemical bonding, ions may also be absorbed by electrostatic forces at the surface, especially if there are already ions present on the surface. (Memming, 1998)

Band Alignment

When the semiconductor and the electrolyte are put in contact, electrons and holes would flow until an equilibrium is achieved. This charge transfer results in an electric field, whose potential balances the initial electrochemical potential differences. When equilibrium is reached, thus, the Fermi energy level is constant throughout the semiconductor and electrolyte. However, when illumination is applied, the energy levels suffer the quasi Fermi level splitting, resulting in a potential difference that would create a constant flow of electrons and result in the redox reaction. This is illustrated in Figure 21, where the figure in the left is before equilibrium, the center shows the band bending at the interface when the system reaches equilibrium, and finally the right one shows how illumination (or applying a voltage) splits the Fermi level and reduces the band bending at the semiconductor surface. (Walter, et al., 2010)

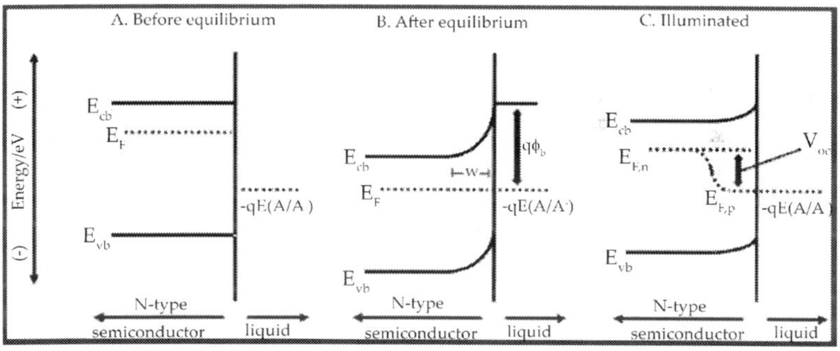

Figure 21: The band energetics of a semiconductor/liquid contact are shown in three cases: (A) before equilibration between the two phases; (B) after equilibration, but in the dark; and (C) in quasi-static equilibrium under steady state illumination. From (Walter, et al., 2010)

The overpotentials required to drive the water splitting reaction are specially affected by the placement of the valence and conduction band edge relative to the redox potentials. (Walter, et al., 2010) If the valence band is not positive enough, or the conduction band is not negative enough, can result in no or little reaction yields. However, in order for both bands to be properly aligned the bandgap of the semiconductor must be rather big, but that also implies that they would not absorb as much light. That is why it often two semiconductors are considered to be connected in series. In general, n-type materials as used as photoanodes and p-type materials are used as photocathodes, since the band alignment and donor and acceptor nature of these materials, respectively, are better suited for these specific functions. (Walter, et al., 2010)

The band alignments of different common semiconductor materials with respect to the water splitting potentials are depicted in Figure 22. By looking at this graph, the potential of a certain material to be used as photoanode or photocathode can be inferred. If the water splitting potential lies within the CB and the VB, then the reaction is favorable, namely 0 V and 1.23 V. However, if they are going to be used only as either simply photocathode or photoanode, only one of these edges matter, 1.23 V for photoanodes such as Fe_2O_3 or WO_3, and to 0 V for photocathodes such as Cu_2O or Si. In addition, in this particular graph, also the potential for CO_2 reduction is included, since there is an increasing trend to research this reaction to create other solar fuels based on carbon. (Xie, et al., 2016)

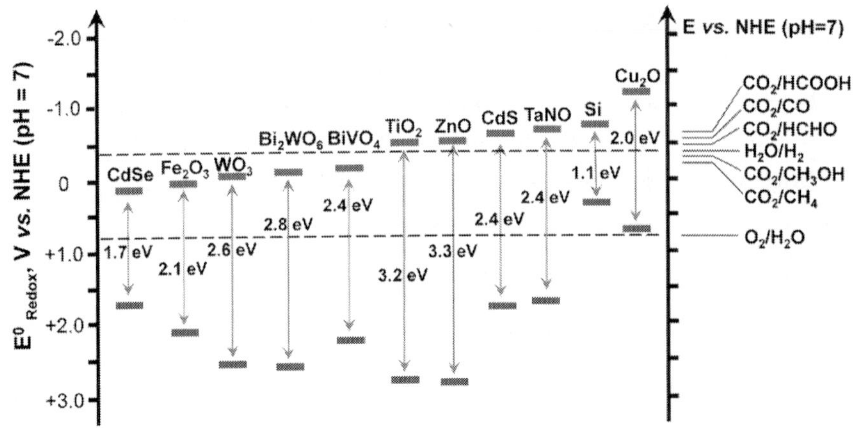

Figure 22. Band edge positions of some typical semiconductor photoelectrode materials relative to the energy levels of the redox couples both involved in water electrolysis and CO_2 reduction. From (Xie, et al., 2016)

The junction between the semiconductor, either p or n, with the electrolyte would result in an difference in charge concentration, creating a depletion region in the semiconductor. (Gryse, et al., 1975) In the electrolyte, this region next to the semiconductor is called the Helmholtz layer, and it is a region where the concentration of certain species is lower than in the bulk of the electrolyte because they are being consumed within the reaction. (Walter, et al., 2010)

The maximum internal energy that can be extracted from such a system is given by the barrier height. This difference should be as large as possible to achieve the best performances. Thus, by tuning the redox potential by changing the conditions of the electrolyte can also be a method to optimize the interface barrier height. (Walter, et al., 2010) These conditions can be tuned base on the Nernst equation shown earlier, which incorporates the temperature and pH. However, this gives only a small room for improvement, and in particular the pH changes have proven largely unsuccessful since the surface groups would protonate or deprotonate to adjust the pH, generating a potential drop and a band shifting. (Hernandez-Pagan, et al., 2012) Thus, the semiconductor has to be carefully chosen in order to achieve optimized semiconductor/electrolyte interface energy band alignment.

First, it is important to understand how the electrolyte polarization affects the properties of the electrode. By doing this, the semiconductor can then be changed to optimize the semiconductor/electrolyte interface. An approach that has been successfully implemented is the tuning of the band edge position by including dipoles ate the semiconductor surface. These dipoles must be then held very close to the semiconductor surface, as illustrated in Figure 23. (Walter, et al., 2010)

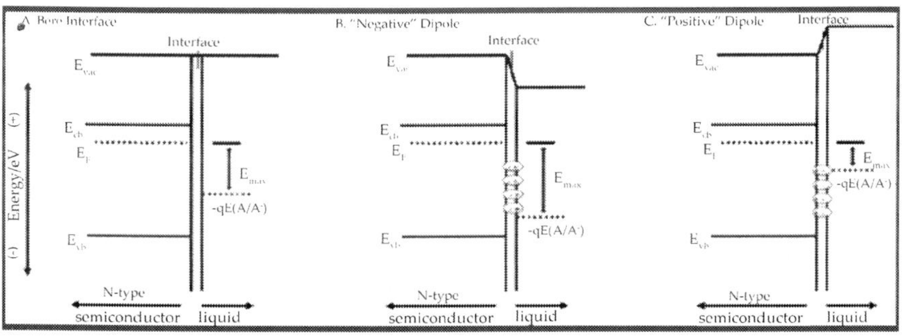

Figure 23. Band diagrams of the semiconductor/electrolyte interface A) bare, B) with a negative dipole and C) with a positive dipole. The presence of an interfacial dipole layer in figures B and C can significantly affect the electrostatic barrier and hence the photoelectrode performance. For simplicity, the band

diagrams are shown at the flat-band condition in the absence of band bending. From (Walter, et al., 2010)

There are several factors that would further affect the semiconductor performance, and will be explored in this section in more detail. First, there would be recombination at the studied interface, which would reduce the charge carriers available. This will be reflected in the open circuit voltage that can be obtained from the semiconductor, also called the onset potential, as well as the fat bland potential.

Recombination Mechanisms

An important factor to take into account when considering this interface are the recombination mechanisms that can occur. Adding to the ones described for the bulk and surface of the semiconductor, further recombination mechanisms occur at the interface. These are associated with the kinetic processes occurring for the redox reactions. The two most important to consider are the tunneling currents and the electron-transport currents associated with majority charge carriers traversing the interfacial barrier. Thus, the charge carrier collection at the interface is a crucial process to ensure an efficient transfer of the electrons and holes into the redox couple. Figure 24 shows the main recombination mechanisms within the band diagram of the system. The electron-hole pairs can thus be recombined through a current density associated with radiative or non-radiative recombination in the bulk of the semiconductor (J_{br}), depletion-region recombination (J_{dr}), and surface recombination due to defects (J_{ss}), tunneling current density (J_t), and electron-transfer current density associated with majority charge carriers traversing the interfacial barrier (J_{et}). In good quality semiconductors such as silicon, these two mechanisms will be likely the predominant recombination mechanisms, and thus becomes especially important to understand these mechanisms. (Walter, et al., 2010)

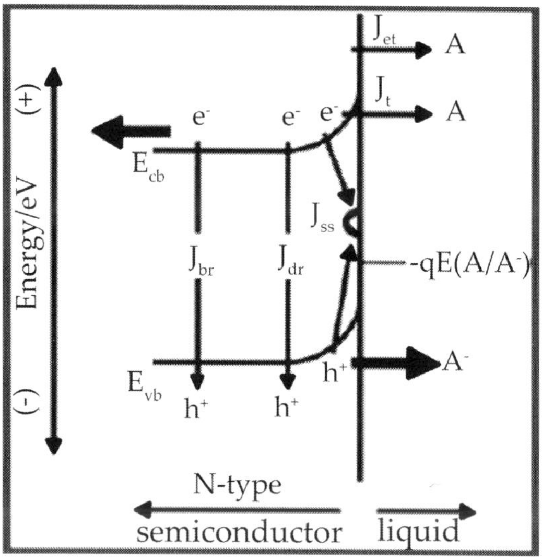

Figure 24: Recombination pathways for photoexcited carriers in a semiconductor photoelectrochemical cell held at open circuit can be broken down into at least five different categories, represented by the thin arrows in the diagram. The electron-hole pairs can recombine through a current density associated with radiative or nonradiative recombination in the bulk of the semiconductor (J_{br}), depletion-region recombination (J_{dr}), surface recombination due to defects (J_{ss}), tunneling current (J_t), and electron-transfer current associated with majority carriers traversing the interfacial barrier (J_{et}). Electron collection by the back contact and hole collection by the redox couple (e.g., oxidation of water to O_2) are processes that contribute positively to device efficiency, and these are depicted by thick black arrows. From (Walter, et al., 2010)

Open circuit voltage

This is a concept borrowed from the PV field, and it is defined as the difference between the quasi-Fermi levels, and can be measured experimentally, also called often onset potential. It is one of the main characteristics that would define if the reaction would happen, aside from the band alignment. This value needs to be higher than 1.23 V, in order to take into account the thermodynamical potential of the redox reaction and any extra overpotentials associated with losses that might occur in the system, often associated with the kinetics of the reaction in a microscopic level. (Walter, et al., 2010) The open circuit voltage would be affected by the recombination mechanisms, both inside the semiconductor and at the interface with the electrolyte. In fact, these recombination mechanisms explain to certain extent the difference between the bandgap and the open circuit voltage. (Gratzel, 2003) Thus, controlling these recombination mechanisms is crucial to achieve good water splitting. In addition, this photovoltage depends on the illumination conditions.

Flatband Potential

This potential is achieved when the Fermi level is constant throughout the whole system, as previously explained. This means that the band bending at the semiconductor/electrolyte interface disappears, as shown in Figure 25. The potential that needs to be applied to achieve these conditions in the flatband potential. The bigger the flatband potential is, the bigger the band bending would be under normal conditions. (Memming, 1998) This is important for the performance of the device because they define the barrier that would form at the interface for charge carrier injection.

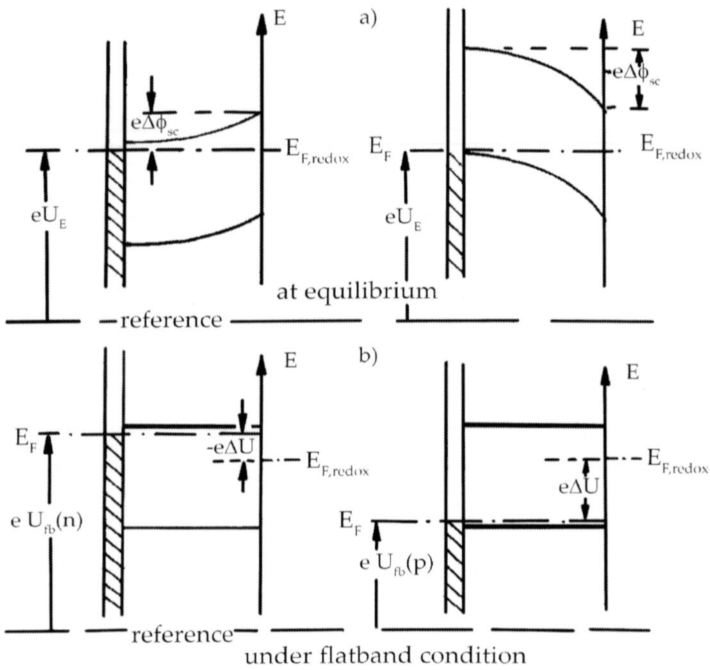

Figure 25: Energy scheme of the n- (left) and p-type (right) semiconductor/electrolyte interface a) at equilibrium and b) under flat band conditions. From (Memming, 1998)

This can also be seen in the energy diagrams shown in Figure 26, taking an n-type semiconductor as an example. Since the band alignment does not have a perfect alignment, there is a slight band bending under equilibrium, which is increased when applying an anodic bias. However, when the bias is cathodic, flat band conditions can be achieved. (Memming, 1998)

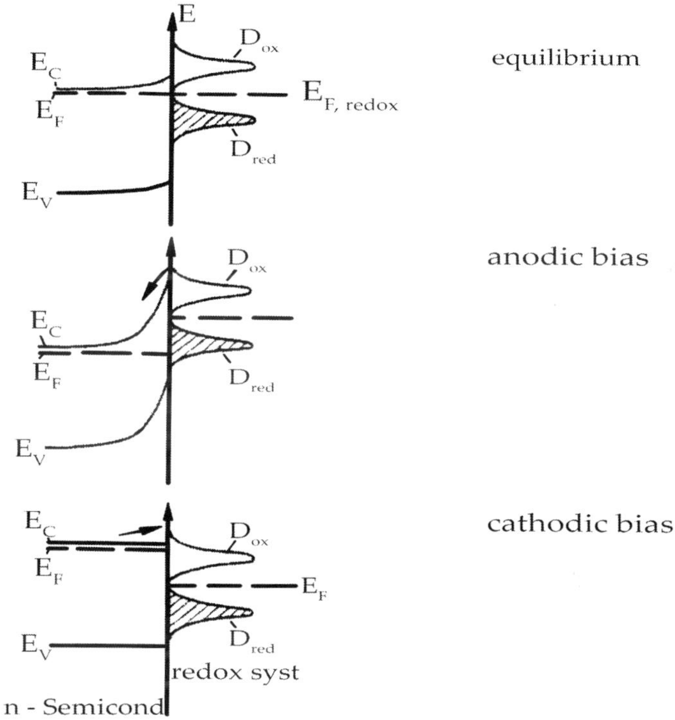

Figure 26. Energy diagram for the interface n-type semiconductor-redox system for different biases, namely at equilibrium (left), with anodic bias (center) and with cathodic bias (right). From (Memming, 1998)

This quantity can be obtained by looking at the Mott-Schottky plots, where the slope would give the donor or acceptor concentration, and the intersection with the x-axis (infinite capacitance) would give the flatband potential. This method will be explained further in following sections. (Memming, 1998)

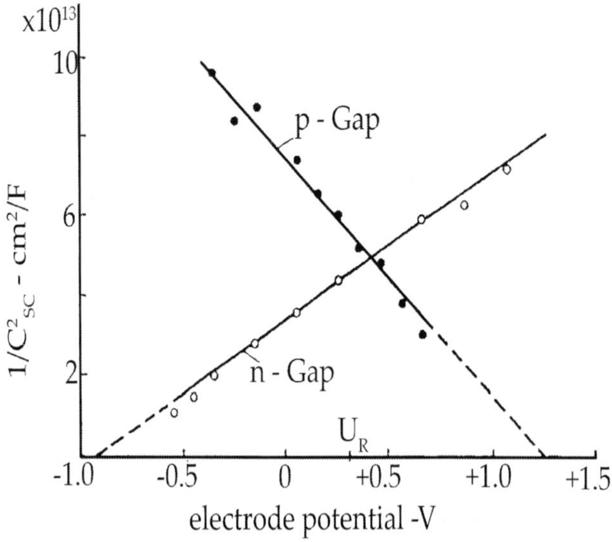

Figure 27: Mott-Schottky plots of the space charge capacity vs. electrode potential for n- and p-type GaP electrodes in 0.1 M H2SO4. From (Memming, 1998)

However, determining the flatband potential is a complex process, and thus this value must be taken with caution. This value depends on the electrolyte, thickness and layers involved in the photoelectrode. More information on this topic will be given when describing the measuring method.

Charge Accumulation at the Surface

As previously commented, when a semiconductor or metal is placed in an electrolyte, a certain charge accumulation appears, creating a double layer at the interface, also called Helmholtz layer. This creates a potential difference at the interface, which affects the rate and mechanisms of the reaction. Thus, it is important to have a clear idea of what this layer is, how it forms and how does it affect the overall system.

Helmholtz Layer

Assuming that the cations are specifically adsorbed on a certain surface, equivalent negative counter charges would accumulate just next to the surface. This is because the cations would react at the surface, leaving the negative ions behind. This charge separation creates a potential difference across the interface, depending on the adsorbed ions, as illustrated in Figure 28. (Memming, 1998) Note that this layer is depicted on the surface of a metal rather than a semiconductor for easiness of depiction, but this effect happens in all surfaces where a reaction happens. The only difference is related to the charge distribution at the metal.

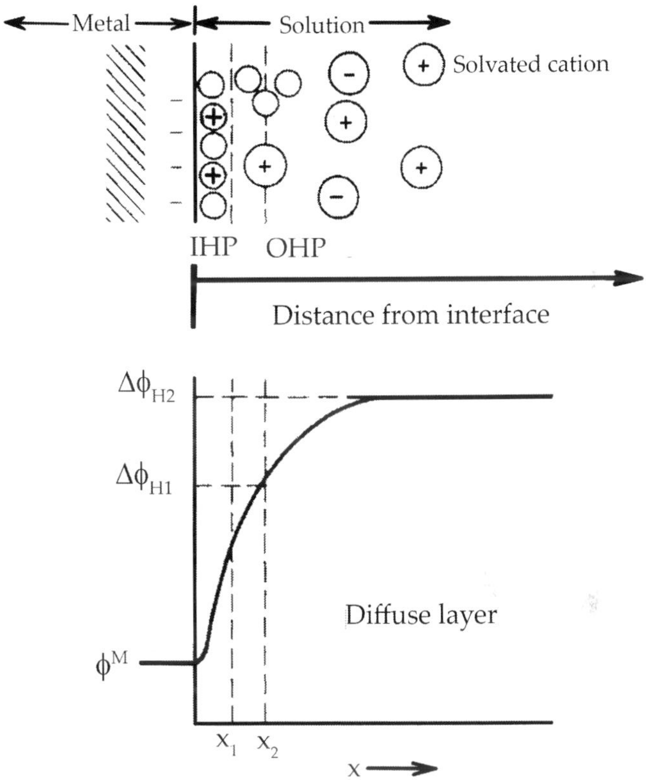

Figure 28: Potential distribution at the metal-electrolyte interface. From (Memming, 1998)

The variations seen in this interface would lead to a change in the potential across the Helmholtz layer. This can be convenient to be able to control it by changing or functionalizing the surface, since this would change the voltage drop caused at the interface, and thus, the overpotentials needed. The characteristics of the Helmholtz layer can be characterized by the Helmholtz capacity, which is related to the total interface capacity as

$$\frac{1}{C_{H,d}} = \frac{1}{C_H} + \frac{1}{C_d}\,. \qquad \text{Equation 28}$$

However, the potential drop across the double layer can be neglected at high concentrations in the case of metals, since the relative concentration differences would not be as pronounced due to the high conductivity of metals. (Memming, 1998)

Nevertheless, if semiconductors are considered, the results and conditions can change, varying the resulting potential drop. In semiconductor materials, the charge carrier density is much smaller than in a metal, and thus this effect would be much more pronounced. The resulting charge distribution is depicted in Figure 29, where the electrolyte is considered to have a high ion concentration, and thus the diffuse layer in that side is neglected. (Memming, 1998)

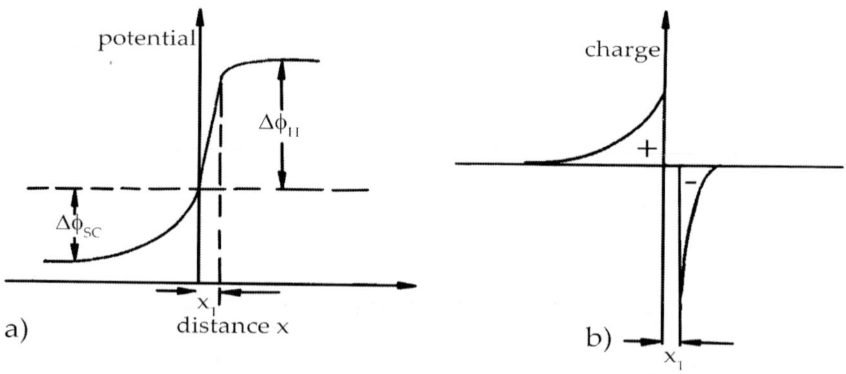

Figure 29. Potential and charge distribution at the semiconductor/electrolyte interface. From (Memming, 1998)

The accumulation of charge at the semiconductor surface causes a potential different and band bending at the interface. This can also be seen when measuring the impedance and capacitance of the system. These are illustrated in Figure 30, underlying the differences between n and intrinsic materials, and how they affect the space charge capacitance.

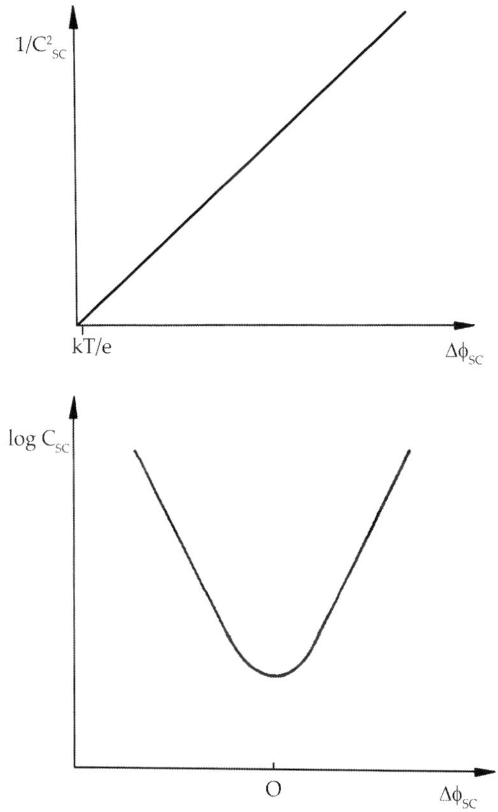

Figure 30. Space charge capacitance vs. potential across the space charge layer in an n-type semiconductor (left) and an intrinsic semiconductor (right). Note that these curves have been theoretically calculated. From (Memming, 1998)

To better understand these values, the math behind the space charge layers In the case of the semiconductor, the total capacitance at the interface can then be defined by the Helmholtz capacitance and the space charge region capacitance, as

$$\frac{1}{C} = \frac{1}{C_H} + \frac{1}{C_{sc}} \quad \text{Equation 29}$$

In addition, it must be taken into account that, especially in the case of semiconductors, defects may appear at the interface, which would increase the surface states and thus the charge carriers accumulated in them. This would change the charge accumulation, and in turn also the

Fermi level positions, as already illustrated in Figure 30. (Memming, 1998)

Band Pinning

The pinning of the bands occurs if the potential across the Helmholtz layer remains constant. Thus, this would require some voltage applied to achieve the status of flatband potential. It must be indicated that the flatband potential of many systems varies with the pH of the electrolyte, with an increase of pH leading to a cathodic shift, and so the energy bands also shift. However, the redox system species do not define the band position of the semiconductor. Thus, the semiconductor properties can be defined only by the material and the working pH. (Memming, 1998)

When band pinning occurs, band bending will be caused at changing illumination and electrical bias. Figure 31 depicts this situation, where the oxidation and reduction potentials are pinned on the surface, and then the semiconductor suffers from band bending. This band bending then causes the holes to be favored over electrons to reach the surface, creating in a way a selective contact at the surface.

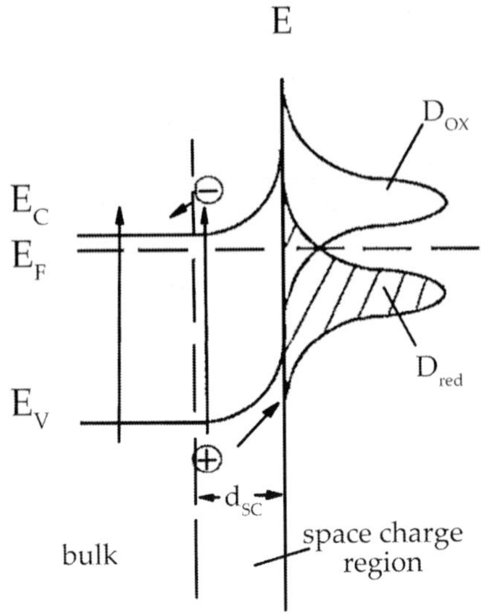

Figure 31: Excitation of electron-hole pairs by light and separation of charge carriers by the field across the space charge layer. From (Memming, 1998)

The band bending will also change depending on the concentration of species in the solution, which it is partly what the pH represents. For example, depending on the pH, either H^+ (acid) or OH^- (basic) ions would be found in the electrolyte. Figure 32 shows how this might occur, by varying the states concentration at the surface. It can then be seen that the minority species control the band bending.

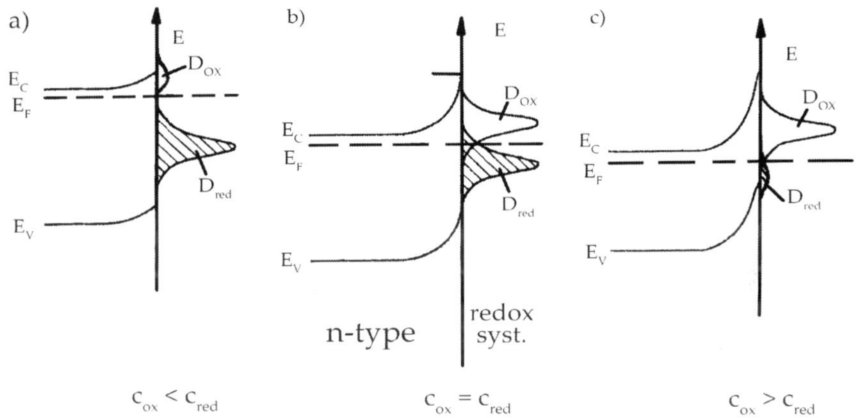

Figure 32: Energy diagram for the semiconductor/electrolyte interface at equilibrium for different concentrations. From (Memming, 1998)

This can be an undesired effect, especially if both half reactions are desired to be conducted simultaneously. However, in some cases this can be changed by modifying the surface chemistry by strategies such as doping profiles, resulting in an unpinning of the bands. In addition, semiconductors can also suffer an unpinning of the energy bands during illumination. The shift is usually cathodic for photocathodes and anodic for photoanodes, with a value of a few hundred of millivolts. (Meissner, et al., 1988) This effect could be explained by trapping of minority charge carriers in surface states, which would act as an additional capacitor and change the potential distribution and band position. (Memming, 1998) The effect of this band pinning an unpinning is better illustrated in Figure 33, taking p GaAs as an example.

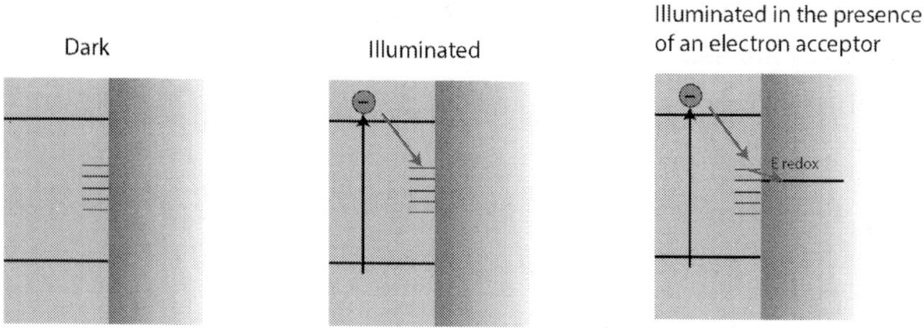

Figure 33. Position of energy bands at the surface of p GaAs in the dark and under illumination. Adapted from (Memming, 1998)

The difference in charge accumulation at the surface can also be described mathematically as

$$\Delta Q_{SS} = C_H\left[U_{fb}(light) - U_{fb}(dark)\right] = C_H\left[\Delta(\Delta\phi_H)\right]_{\delta}$$

Equation 30

where ΔQ_{SS} is the charge difference at the surface. These mathematical interpretations can also be translated into Mott-Schottky plots, such as the one depicted in Figure 34, where the capacitance caused at the space charge regn is depicted for dark and illuminated conditions, showing the higher accumulation when light is producing charge carriers.

$$\Delta Q_{ss} = C_H\left[U_{fb}(\text{light}) - U_{fb}(\text{dark})\right] = C_H\left[\Delta(\Delta\phi_H)\right]_{h\nu} \quad (5.49)$$

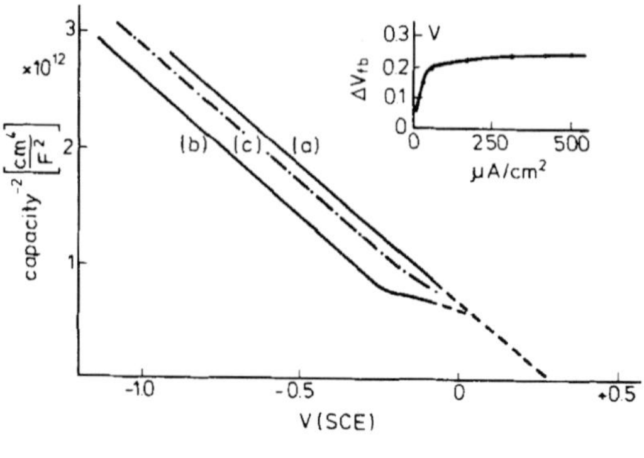

Figure 34. Mott-Schottky plot of the space charge capacity vs. electrode potential for p-GAs at pH 1, in the dark and under illumination. Inset: Flat-band potential vs. limiting photocurrent. From (Memming, 1998)

It is interesting that this shift in the Mott-Schottky plot between dark and illuminated conditions already occurs at low illuminations, and saturates at higher light intensities, since at that point all the surface states would be occupied by charge carriers. This saturation is especially evident when looking at the flat band potential changes, where the curve saturates after low illuminations.

If band pinning is desirable, it must be noted that this effect of band unpinning can be avoided by introducing a suitable redox system in the electrolyte, such as adding an electron acceptor molecule to the solution when there is a p-type semiconductor as a photoelectrode. In that case, the trapped steps would be further transferred to the solution.

Catalysis

Even when a semiconductor material has the right band alignment and bandgap to drive the water splitting, the kinetics of the reaction on the semiconductor surface can often limit the reaction rate. This can be overcome by an overpotential. However, this overpotential comes at a cost. Another strategy is thus to use a catalyst at the surface to improve the kinetics of the reaction. (Walter, et al., 2010) Figure 35 shows the effect of the catalyst on the energy diagrams, showing how intermediate energy states occur at the surface close to the redox reaction potential, helping the charge carrier participation into the redox reaction.

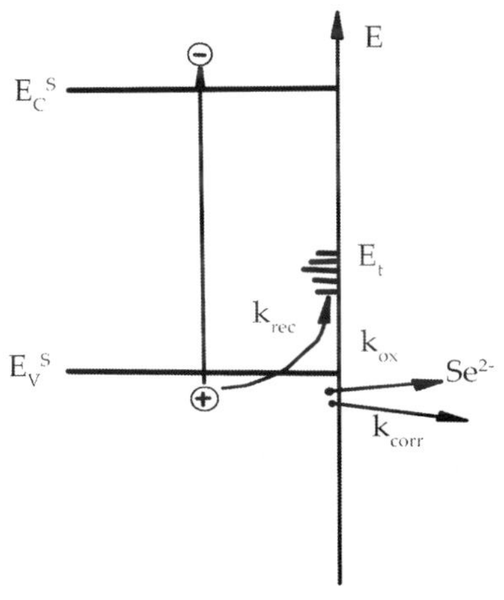

Figure 35. Oxidation of a redox system and anodic corrosion by hole transfer at an n-type electrode during light excitation. From (Memming, 1998)

This is often applied as nanoparticles, in order for the film to appear transparent and not affect the optical properties of the electrode, while at the same time not creating a Schottky barrier at the metal/semiconductor interface. If a Schottky barrier is formed, then it would equilibrate the Fermi level with the H^+/H_2 redox potential, moving the dependence of the barrier height to that of the semiconductor/metal junction. In addition, when the barrier height is inhomogeneous, as in the case of the catalyst nanoparticles, the junction exhibits the so-called Fermi level pinning, in which the barrier is reduced. (Walter, et al., 2010) Thus, there is a balance stroked between charge carrier injection via the catalyst and charge carrier recombination at the surface states. (Memming, 1998)

The quality of the catalytic properties can be determined by the needed voltage that would drive the reaction. It has been established that a good catalysts needs approximately 100 mV to drive a current density of 10 mA/cm^2, while a bad catalyst would need at least 600 mV for the same current density. Note that these values only refer at the kinetic characteristics of the reaction, and the open circuit voltage should not be affected, since the thermodynamic characteristics would remain the same. These thermodynamic effects are reflected on the fill factor of the photoelectrode, meaning the "squareness of the JV characteristic." (Walter, et al., 2010) The meaning of these properties is more clearly illustrated in Figure 36.

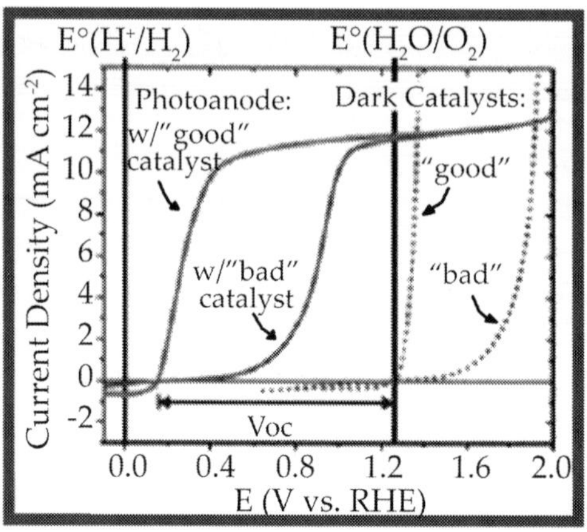

Figure 36. Schematic showing the qualitative effect of surface catalysts on photoanode performance. Dotted curves show the current density vs overpotential relationship for generic "good" and "bad" catalysts. The solid curves show the

performance of a hypothetical oxygen-evolving semiconductor photoanode integrated with either a “bad” or “good” catalyst. From (Walter, et al., 2010)

In summary, a good catalyst for water splitting must satisfy two basic requirements: it must be active towards the water splitting reaction, producing large currents at low overpotentials, and must be robust enough to maintain a stable performance throughout its lifetime. Although many commercial catalysts for industrial electrolysis satisfy these requirements, further improvements could be made. In addition, the materials used are often considered to be rare and expensive. (Lettenmeier, et al., 2015) To determine which catalysts could be used, the Tafel relationship could be used (Maruthapandian, et al., 2017), as

$$i = i_0 exp\left(\frac{\alpha_a F\eta}{RT}\right).$$

where i0 is the exchange current (forward and reverse reactions at equilibrium), αa is the transfer coefficient for the anodic reaction, η is the overpotential, F is the Faraday constant, R is the ideal gas constant and T is the temperature. (Doyle & Lyons, 2013) The dependence of these relations is exemplified in figure 37.

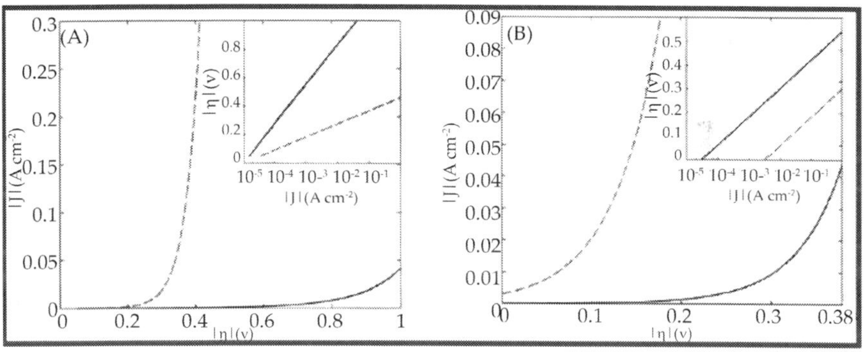

Figure 37. Polarization curves and Tafel plots (inset) illustrating the electronic and geometrical effects of catalysts. (A) Dashed blue: J0) 10^{-5} A cm^{-2}, b) 40 mV/decade. Solid black: J_0) 10^{-5} A cm^{-2}, b) 120 mV/decade. (B) Dashed blue: J0) 3 × 10^{-3} A cm^{-2}, b) 120 mV/decade. Solid black: J_0) 3 × 10^{-5} A cm^{-2}, b) 120 mV/decade. From (Walter, et al., 2010)

To determine how good a catalyst can be for the water splitting reaction as an anode or a cathode the so-called volcano plots can be fabricated, where the activity of a catalyst is measured depending on the M-H bonds between the catalyst and the hydrogen atoms. These plots show that the HER activity would rise up to a certain point, after which in decreases with increasing bond energy, as displayed in Figure 38. Although the exact conditions such as pH or temperature would affect these values, the general trend remains the same. (Walter, et al., 2010) It must be noted that, both in acidic and base solutions Pt and Pd demonstrate exceptional activity. However, they are rather precious and expensive metals. Thus, other materials such as Ni can be taken into account. Further exploration of the specific materials and properties of the catalysts will be explored in following sections.

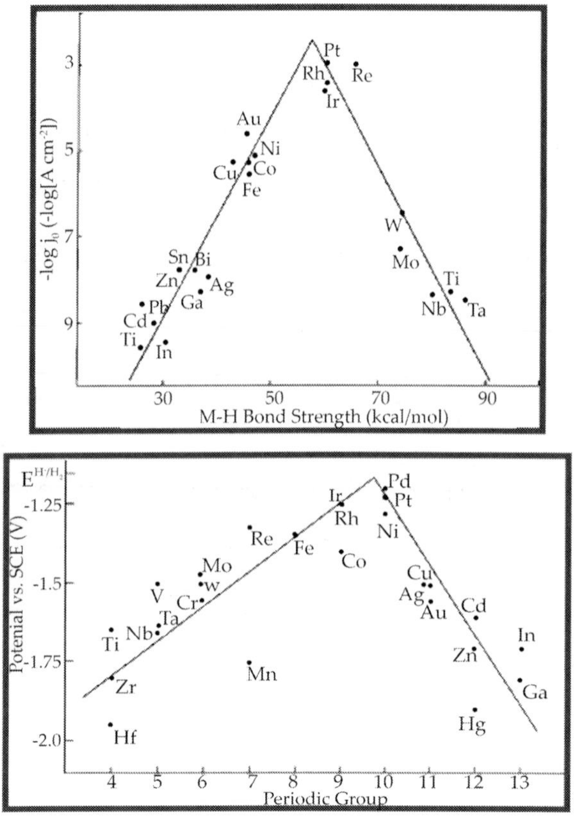

Figure 38. The volcano relation for pure metals in acidic solution (left) and in 30 wt. % base solution. The noble metals Pt and Pd demonstrate exceptionally high activity, with Ni as the most active non-precious metal. From (Walter, et al., 2010)

EFFICIENCY DEFINITIONS

The efficiency of a solar hydrogen production system is usually measured by the overall solar-to-hydrogen (STH) efficiency. This is the most important and reliable parameter that can be used to compare different systems. However, the definition of efficiency is not always clear. (Chen, et al., 2013) Thus, it is important to clearly define some guidelines to allow for reliable comparison between different solar water splitting systems. (Chen, et al., 2013) proposed different valid methods to define efficiency, as well as identified their different pitfalls.

Solar-to-hydrogen (STH) efficiency.

This is the most common one, since it describes the overall efficiency of the system exposed to solar irradiation (AM1.5) under zero bias conditions, meaning that all the energy produced for the reaction comes from the sun, with no other additional electrical energy applied. This gives an idea of how the system would perform under normal operation conditions. However, for this measurement to be effective, the device must be stand-alone, and no additional voltage must come from external sources of power. Thus, this efficiency definition cannot be applied to many single photoelectrode systems, since they are often not able to cover the necessary voltage on their own.

Moreover, it is important to point out here that, for this measurement to be reliable, both electrodes should be immersed in the same solution, and not two different compartmentalized solutions with two different conditions. In addition, there should be no sacrificial donors or acceptors in the reaction, meaning that the electrode should stay the same throughout the reaction. Then, the STH efficiency can be calculated as the chemical energy of the hydrogen produced, which can be calculated by using the hydrogen production and the change of Gibbs free energy previously described, since it describe the maximum amount of energy that can be recovered from the hydrogen molecules. In addition, the area comes into play, and thus it should be accurately measured and defined as the illuminated area. (Chen, et al., 2013)

$$\eta_{STH} = \left[\frac{\dot{n}_{H_2}\left(mmol \frac{H_2}{s} \right) x \Delta H \left(237 J / mol \right)}{P_{total}\left(\frac{mW}{cm^2} \right) x\, Area\left(cm^2 \right)} \right]_{AM1.5G} \quad \text{Equation 31}$$

where $\dot{n}_{H_2}$. is the molar flow of hydrogen, ΔH is the energy of the reaction and Ptotal is the total irradiation falling in the system per unit of area.

Disadvantage of the STH efficiency is that this method uses the measure of the hydrogen produced, which is analytically difficult to determine. Alternatively, the relation between the hydrogen and the photocurrent produced can be related by het Faradaic efficiency, obtaining a much easier expression to use (Chen, et al., 2013)

$$\eta_{STH} = \left[\frac{J_{SC}\left(mA / cm^2 \right) x 1.23 V x \eta_F}{P_{total}\left(\frac{mW}{cm^2} \right)} \right]_{AM1.5G} \quad \text{Equation 32}$$

where the voltage of 1.23 V represents the thermodynamical voltage of the reaction previously discussed, Jsc is the photocurrent produced when the applied voltage is zero and η_F is the faradaic efficiency. It is important to note that these equations are only valid when the production of H_2 and O2 is stoichiometric, and there are no addition al sacrificial electrodes or secondary reactions occurring. In particular, the second definition assumes that the measured photocurrent solely contributes to the water splitting reaction. If that is not the case, the resulting efficiency would be overestimated. (Chen, et al., 2013) Thus, other methods to define the efficiency of the system can be envisioned to complement this one.

Applied Bias Photon-to-Current Efficiency (ABPE)

This efficiency is used more as diagnosis tool to explain observed behaviors than to define the effectiveness of the system as a whole. In this case, a bias voltage is applied between working and counter electrode, increasing the current produced in the device. Mathematically, it can be described as

$$ABPE = \left[\frac{J_{ph}\left(mA/cm^2\right)x\left(1.23V-|V_b|\right)x\eta_F}{P_{total}\left(\frac{mW}{cm^2}\right)}\right]_{AM1.5G} \quad \text{Equation 33}$$

where J_{ph} is the photocurrent density obtained under an applied bias Vb.

It is important to realize that this method cannot be used as a reliable performance standard, since by applying a voltage the system becomes closer to standard electrolysis in dark conditions. (Chen, et al., 2013) In addition, when reporting ABPE, it must be clearly stated a voltage was applied. It is often the case that the voltage between the working electrode and a reference electrode is reported as zero, ignoring the voltage that was needed between the reference electrode and the counter electrode. In addition, as in the previous case, sacrificial electrodes and chemical potentials such as differences in pH must be avoided. Thus, this is an important tool to diagnose the potential and issues with the working electrode, separating them from the ones for the counter electrode, but cannot be directly translated into STH efficiencies. (Chen, et al., 2013)

Incident Photon-to-current Efficiency (IPCE).

This is another diagnosis tool, in which the photocurrent produced by incident photon is described. This measurement occurs through a wavelength range, allowing estimating the maximum STH efficiency possible for a certain photoelectrode, especially when performed under no bias. The mathematical definition of this measurement is defined as

$$IPCE = \frac{electrons/s}{photons/s} = \frac{J/q}{P/\left(hc/\lambda\right)} \quad \text{Equation 34}$$

where J is the total current density produced P is the total light irradiation at a certain wavelength and hc/λ is the energy per photon.

This measurement is a very efficient tool to determine issues in the materials, being able to identify limits in the photon absorbance, the charge transport to the semiconductor/electrolyte interface, and the efficiency of the interfacial charge injection. However, this method assumes that the counter electrode is not limiting to the current. (Chen, et al., 2013)

A variation of this method is the absorbed photon-to-current efficiency (APCE), in which the possible optical losses are extracted in order to understand the inherent performance of the material. Thus, in this case the IPCE measurements are normalized by the absorption, giving more information about the charge carrier kinetics and reaction kinetics of the system. (Chen, et al., 2013)

STABILITY CONSIDERATIONS

Along with the reaction rate, stability is one of the main factors determining the applicability of a certain solar water splitting system. The stability issues can come from several sources, including the semiconductor, and the catalysts for both the HER and OER.

Anodic Decomposition

Many semiconductors, especially with lower bandgap energies (< 3 eV), suffer from anodic decomposition. (Memming, 1998) This occurs by hole consumption, and thus in p-type semiconductors in the dark and n-type under illumination. Due to their importance of stability for PEC applications, the anodic dissolution of germanium and silicon has been widely investigated. In this mechanism, when electrons or holes are injected, they etch away the material, incorporating the material to the solution. The mechanism of the solution, which needs four ions to occur, is depicted in Figure. (Memming, 1998)

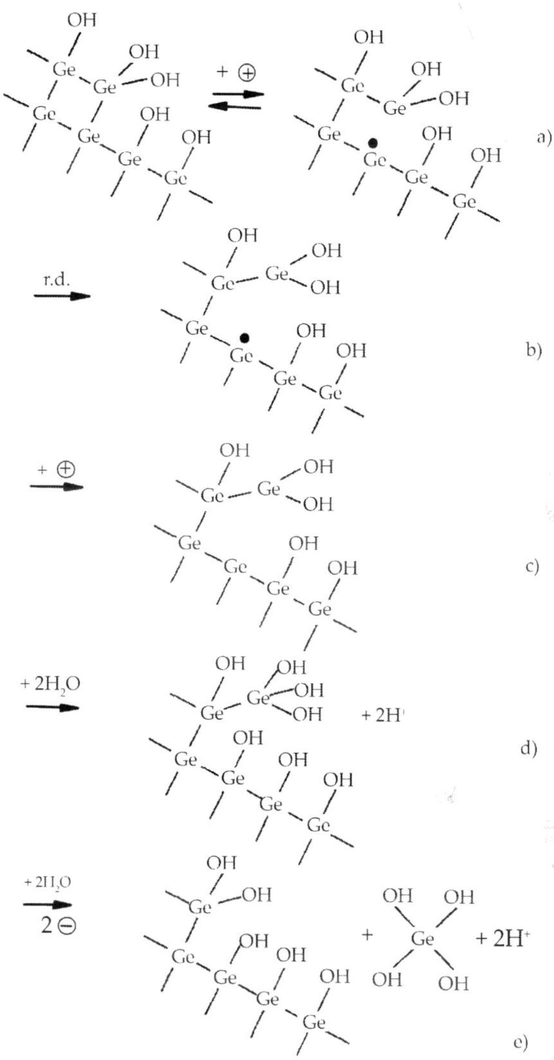

Figure 39. Mechanism of the anodic dissolution of germanium. From (Memming, 1998)

Regarding the energy bands, the situation is depicted in Figure 40. When the charge carrier, a hole in this case, is injected into the electrolyte together with the dissolved Ge, middle gap states are form, which would then lead to recombination of the charge carriers, reducing the efficiency in addition to the drop in stability.

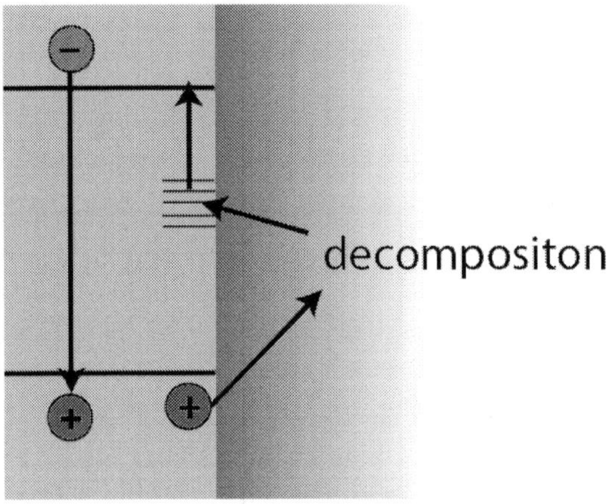

Figure 40. Formation of intermediate states (surface states) during the anodic dissolution of germanium. Adapted from (Memming, 1998)

This degradation can also be detected in the JV curve in the material, especially in dark conditions. The degradation of the photoelectrode would depend in several factors such as the amount of impurities in the material or the doping concentration. (Parkinson, 1984) In addition, the corrosion increases with the resistivity of the semiconductor material, with a linear relation between the anodic limiting current and the ratio of hole density and diffusion length. This is because the saturation current of minority charge carriers will be given by that relation. (Memming, 1998)

As mentioned, the corrosion reaction would then be in competition with the redox reaction of water splitting. It is very important to understand this balance in order to make reliable and durable water splitting systems. This would depend on the thermodynamic and kinetic conditions of each reaction, and the corresponding JV characteristic for each reaction. This can be better understood looking at the energies of each reaction, as shown in Figure 41. If Eredox is lower than Edecomp, the decomposition reaction would dominate. On the other hand, if the Eredox is closer to the valence band, that would be the dominant reaction. (Memming, 1998)

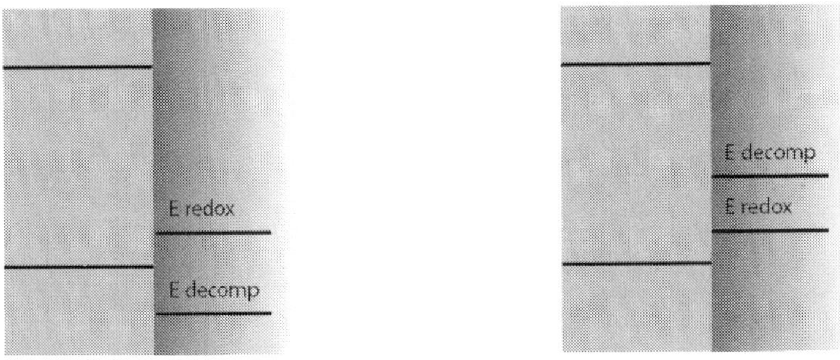

Figure 41. Decomposition and redox potentials with respect to the position of the energy bands. In a) the redox reaction will be favored where in b) the decomposition reaction will be favored. From (Memming, 1998)

Thus, when choosing the correct material for designing a photoanode, not only the band alignment to the redox reaction for water splitting must be taken into account, but also the alignment to the possible decomposition reactions. Note that this alignment can also be tuned by other conditions such as the pH of the electrolyte. Therefore, the whole system has to be considered when trying to make PEC systems more stable.

Cathodic Decomposition

Under cathodic polarization, hydrogen is usually formed when the electrons are transferred from the conduction band into o the electrodes. Usually, in these cases the electrode decomposes leaving a metal layer at the surface, which would influence the electrochemical behavior of the semiconductor electrode. It is also possible that a metal oxide layer remains at the surface, which would then stabilize the electrode for further cycles. The rate of decomposition is then dictated by the redox system, and can be determined by looking at the current densities in dark and illuminated conditions. Some semiconductors such as GaAs would also decompose into the solution if certain compounds such as Br_2 or H_2O_2 are present. (Memming, 1998)

In general, both for the case of anodic and cathodic decomposition, the likelihood of a decomposition reaction can be estimated by the energetic positions of both the energy bands of the semiconductors and the energies of the decomposition reactions, similarly to the anodic decomposition case. The possible situations are depicted in Figure 42. If both anodic and cathodic decomposition energies are within the bandgap, the reduction and oxidation are both thermodynamically possible. If these reaction occur out of the bands, then the semiconductor is considered to be stable.

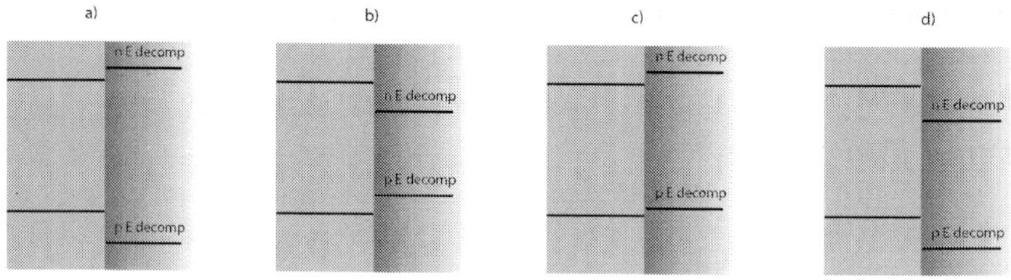

Figure 42. Possible positions of the decomposition energy levels relative to the bandages: a) for a relative stable semiconductor, b) implies instability if either electrons or holes reach the surface, c) and d) imply instability with regard to holes and to electrons, respectively. From (Memming, 1998)

Note that these positions would also depend on other factors such as temperature and solution pH. In addition, although the thermodynamics can be disfavorable, the kinetics of decomposition may be faster, due to other factors such as adsorption, surface chemistry, etc. (Memming, 1998) Thus, although it is useful to look at the band structure to determine the possible decomposition of a semiconductor, it must be considered that it is not the only factor.

Catalyst Stability

There are several factors that can affect the performance of a catalysts. This includes corrosion, catalyst poisoning by solution impurities, and possible changes on the electrode composition and morphology. It is important to take into account that these effects happen in different time scales, some of which are very fast and some are very slow. It must also be taken into account that these mechanisms highly depend on the conditions of the process, including temperature, pH, electrolyte composition, cell housing, current density and potential. (Jong, et al., 1999) Unfortunately, complete avoidance of catalyst degradation is not possible, but it can be slowed down by regulating the system parameters and materials used. Also, precious metal based catalysts such as RuO_2 or Pt tend to suffer less degradation. Finally, adding inert materials such as metal oxides (SnO_2, TiO_2) can help stabilize the electrode and the catalyst. (Chen, et al., 2013)

SUMMARY

PEC devices are often composed of semiconductors, which are materials with bandgaps in between that of metals and insulators, meaning that charge carriers can be generated if enough energy to surpass that of the bandgap is given. There are several factors that make a material a good semiconductor for PEC water splitting.

The first factor that must be taken into account is that the water splitting reaction as a thermodynamic potential of 1.23 V, meaning that the semiconductor chosen should in principle produce that voltage upon illumination. However, due to losses in the system, such as entropic losses, cabling, charge transport and injection losses, the practical voltage needed is somewhere between 1.6 to 2 V. In order to create such voltage, a considerably high bandgap would be needed. That would also imply that the photons that the semiconductor would be able to absorb is limited to the blue and ultraviolet light, reducing the amount of charge carriers produced, and thus the performance of the device. Thus, a trade-off is established. Alternative approaches can also ease this limitation, as would be seen in further chapters.

Another important factor is related to the quality of the material. Once the charge carriers are generated, they need to be transported and collected at the interfaces. However, if the material has a high amount of defects, theses charge carriers could recombine before reaching the interfaces, liberating their energy and reducing the amount of charge carriers available to drive the PEC reaction. Thus, high quality and pure materials are needed to assure high performance. Unfortunately, these materials also tend to be rather expensive due to the demanding processes needed to reach such high quality materials, and thus another tradeoff is established for the design of these devices.

The third important point to consider when designing a PEC device is the interface between the semiconductor and the electrolyte. The conduction and valence band alignment of the semiconductor with the redox reaction would determine how efficiently the charge carriers can be injected into the electrolyte and drive the redox reaction of water splitting. This depends mostly on the material properties and that of the electrolyte such as the pH. In addition, depending on the band alignment, the semiconductor material can be used as a photoanode, a photocathode or both. Moreover, the band alignment would also play a role on the stability of the device, since decomposition reactions also have an energy level. If this energy level aligns better with the semiconductor bands than that of the redox reaction, that reaction would be favored, seriously affecting the performance and durability of the PEC device. Thus, it is fundamental to consider the band alignment for the performance and durability of the PEC device.

In following chapters, these three components would be investigated, and possible materials that can cover all the requirements would be suggested. Moreover, more advanced techniques can be applied to tackle some of these challenges, and thus it is worth mention and explores them further in order to achieve a feasible PEC device.

CHAPTER 3

MATERIALS FOR PHOTOELECTRODES

In previous sections, several characteristics have been labelled important when looking at the PEC performance of a photoelectrode. First, the bandgap and other optical properties would affect the light absorbed, and thus, the possible current density produced. In addition, the quality of a suitable material must be good enough for charges to travel through without recombining in the possible defects and trap states. Finally, the band alignment and stability would determine the chemical performance of the semiconductor as a photoelectrode for water splitting. The roadmap for PEC determines that these devices should reach a STH efficiency of 10 % or more to be feasible (Miller, et al., 2010). There are several candidates that can tackle many of these challenges. However, there is no perfect material that can fulfill all. Thus, improvements regarding these materials must be achieved to reach the established targets. In this chapter, a general introduction to the most common semiconductors and their structures will be given, followed by the most common semiconductors used specifically for PEC, and some stability considerations when choosing the semiconductor to be used in a PEC device. In addition, the semiconductor is not the only material present in a PEC device. Here, account of other materials needed such as protective layers and catalysts will be given.

SEMICONDUCTOR MATERIALS

Before going in depth into the semiconductors used for PEC, it is important to understand what a semiconductor is and how it behaves. Semiconductors are materials whose electrical properties lie between those of insulators and conductors. To easily illustrate this, the resistivity of these materials can be used. Typical resistivity ranges at room temperature are displayed in Table 1. On one hand, insulators have a very high resistivity. On the other hand, metals have very low resistivity. Semiconductors lie somewhere in the middle.

Table 1: Typical room temperature resistivity ranges for different material types (Balkanski & Wallis, 2000)

Category	ρ (Ω cm)
Insulators	10^{12}
Semiconductors	10^{6}-10^{-3}
Metals	10^{-6}

In addition, unlike the materials previously described, the resistivity of semiconductors tends to be highly dependent on temperature. When temperature increases, the resistivity of semiconductors decreases, unlike in metals where the resistivity weakly increases with increasing temperature (Balkanski & Wallis, 2000). These trends on the resistivity with temperature for metals and semiconductors are displayed in Figure 1.

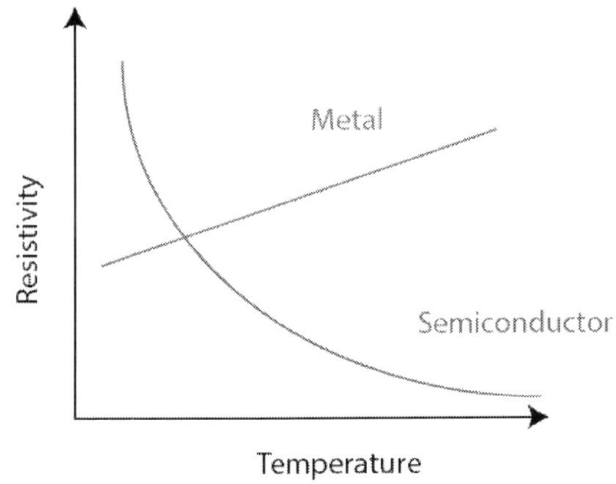

Figure 1. Resistivity versus temperature for a metal (blue) and a semiconductor (red). Adapted from (Balkanski & Wallis, 2000)

This can also be represented by the energy properties inside the atoms. The outermost electrons can have energies either in the valence band, where they are strongly tied to the nucleus, or the conductive band, where the electrons are relatively free. In between these energy levels, there is a series of forbidden levels called band gap energy. (Hybertsen & Louie, 1986) The concept is that the metal, semiconductor and insulator materials can be differentiated by the band gap as already explained in previous sections of this book.

Thus, in the case of metals, very little energy if any is needed to detach an atom from the forces of the nucleus into the conduction band, explaining the high conductivities and low resistivities previously mentioned. On the other hand, the energy needed to liberate an electron from the bonds with the nucleus in an insulator is considered so large that it effectively does not allow for any flow of electrons, and thus the high resistivities. In semiconductor, the bandgap energies lie in the middle, where it is possible to break the bond between outer electrons and atom nuclei, but some energy is needed for this to happen. That would also explain the lower resistivity at higher temperatures, since energy is being introduced in the system. (Balkanski & Wallis, 2000)

Some of the most important semiconductors are silicon (Si) and germanium (Ge). Especially, silicon is widely used because of its abundance on the Earth crust and its good semiconductor properties. Carbon in several forms, such as graphite, can also be a semiconductor.

Other elements that can be semiconductors are selenium (Se) and tellurium (Te). These are mostly from the IV (group 14) and VI (group 16) groups in the periodic table (Balkanski & Wallis, 2000). Overall, the single elements that can be considered semiconductors and their position in the periodic table are highlighted in blue in Figure 2.

1	2	3	4	5	6	7	8	9	10	11	12	13	14	15	16	17	18
H																	He
Li	Be											B	C	N	O	F	Ne
Na	Mg											Al	Si	P	S	Cl	Ar
K	Ca	Sc	Ti	V	Cr	Mn	Fe	Co	Ni	Cu	Zn	Ga	Ge	As	Se	Br	Kr
Rb	Sr	Y	Zr	Nb	Mo	Tc	Ru	Rh	Pd	Ag	Cd	In	Sn	Sb	Te	I	Xe
Cs	Ba	*	Hf	Ta	W	Re	Os	Ir	Pt	Au	Hg	Tl	Pb	Bi	Po	At	Rn
Fr	Ra	#	Rf	Db	Sg	Bh	Hs	Mt	Ds	Rg	Cn	Nh	Fl	Mc	Lv	Ts	Og
Lanthanides		*	La	Ce	Pr	Nd	Pm	Sm	Eu	Gd	Tb	Dy	Ho	Er	Tm	Yb	Lu
Actinides #		Ac	Th	Pa	U	Np	Pu	Am	Cm	Bk	Cf	Es	Fm	Md	No	Lr	

Figure 2: Periodic table, highlighting in blue the semiconductor elements, and in yellow the elements more commonly used to make semiconductor compounds

However, the combination of several elements that are not semiconductors can yield a material with semiconductor properties. Thus, regarding the compounds that can be considered semiconductors, the list is very extensive. Some of the most important are the III-V compounds such as GaAs and InSb; II-VI compounds such as CdS, CdSe and ZnSe; the III-VI compounds such as GaS and InSe; and the IV-VI compounds such as PbS and PbTe. Other important group of binary compounds is metal oxides such as TiO_2, WO_3 or ZnO. There are also ternary compounds that have the characteristics of semiconductors, as chalcopyrite ($CuFeS_2$), for example. These would be especially relevant for the field of PEC water splitting, as it would be further explored in the following sections. Finally, there are also organic semiconductors, such as anthracene ($C_{14}H_{10}$) and polyacetylene $(CH)_x$ (Balkanski & Wallis, 2000). The main elements that are used in these semiconductor compounds are highlighted in yellow in Figure 2.

It must be taken into account that not all these elements and compounds can always be used as a semiconductor. On the contrary, this also depends on the crystal structure of the semiconductor, such as that of carbon as graphite or graphene, or the differences between crystalline silicon and amorphous silicon. Thus, in the next section, these differences will be explored, and it must always be taken into account when choosing a certain material for use as a photoelecrode.

Semiconductor Structures

Semiconductors, as any solid, can either appear as a crystalline or an amorphous structure. Crystalline structures are characterized by its atoms being arranged in a periodic array. On the other hand, amorphous materials lack a long range order (Balkanski & Wallis, 2000). There are also intermediate structures such as poly-crystalline or micro-crystalline materials, in which the structure is composed by several crystal grains instead of one big crystal. A representation of the atom organizations in each material type is depicted in Figure 3.

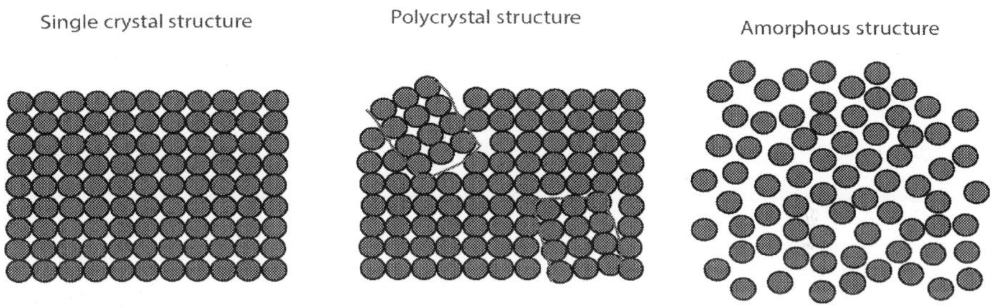

Figure 3. Schematic representation of the atom organization of crystalline (left), polycrystalline (center) and amorphous (right) materials

In addition, also the defects introduced by impurities, doping and other phenomena must be taken into account when considering the internal structure, which in turn would affect the electrical, optical and recombination properties of these materials. These properties will be mainly divided in two groups: the crystalline semiconductors and the amorphous semiconductors. Poly and micro or nano-crystalline semiconductors are assumed to have properties ranging in between these two cases.

Crystalline Semiconductors

The crystal orientation would affect the final properties of the semiconductor device fabricated. In addition, it must be considered that, apart from the orientation of the unit cell, some materials present different crystalline structures depending on the deposition conditions (pressures, temperatures, etc.) and post-deposition treatments. Depending on which crystalline structure is used, different electrical, mechanical and optical properties will be obtained. An example of this transition is the possibility of carbon to be presented as diamond, graphite or graphene, which is a variation of the graphite structure.

Finally, it must be considered that many semiconductor properties can be dramatically changed when defects are introduced in the structures. The defects can usually be classified as impurities, in which case an alien atom either substitutes an atom of the target material or is placed in between the atoms of the target material; voids or vacancies, in which an atom is missing in the structure; interstitial atoms, in which there is an additional atom of the target material that deforms the crystalline structure; and Frenkel defects, in which an atom from the target material has moved from a site in a crystalline matrix to an interstitial site. All these defects are illustrated in Figure 4 for an example of crystalline silicon.

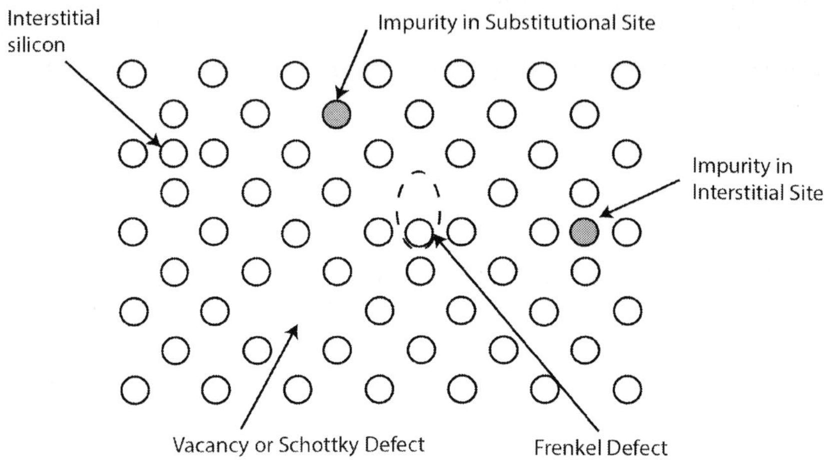

Figure 4. Illustration of different defects that can appear in a crystalline material.

These types of defects in a crystalline structure can ultimately lead to dislocations in the crystal structure. This only occurs when several repetitions of these defects occur in a certain area of the crystal. Such dislocations, which are shown in Figure 4, can create problems like recombination of charge carriers and grain boundaries. In addition, these dislocations can create stresses inside the crystal, which can lead to fracture. Thus, they must always be avoided when processing crystalline semiconductors.

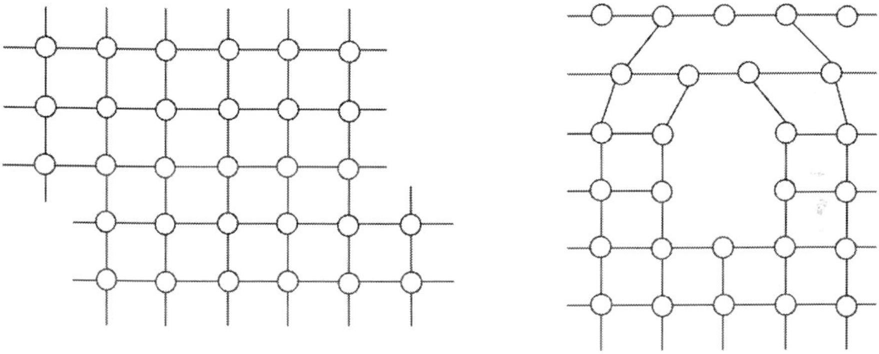

Figure 5. Illustration of Dislocation Defects

These defects usually occur due to high temperature treatments such as the ones used in wafer fabrication. In addition, it can also be caused by thermal gradients created during oxidation diffusion, or other processes that require high temperatures. This is especially prevalent in the annealing processes required to achieve a certain crystalline structure. Thus, precise control of the processing conditions is crucial to achieve a good quality of the semiconductor material, especially regarding crystalline semiconductors.

Another consideration when depositing several different crystalline materials together is the lattice feature size. If the lattice dimensions of the two deposited materials are too different, this will lead to lattice mismatches and defects, i.e. dislocations. (Roxlo, et al., 1994) These considerations are especially relevant for semiconductor devices based on elements of groups III and V of the periodic table, such as GaAs, InP, GaInAs, etc. These devices commonly consist of several layers of different materials, each of them with a different lattice constant. Thus, when combining them, the target would be to have fairly similar lattice constants to avoid dislocations in the structures. (Law, et al., 2010)

Nevertheless, the lattice mismatches are also partially affected by the deposition techniques. Several researches reported the growth of certain semiconductor structures on lattice-mismatched substrates using epitaxial growth combined with an extremely accurate alignment method. (Chuang, et al., 2007) Thus, choosing the right deposition method and accurately controlling the deposition conditions is especially crucial in these cases.

Amorphous Semiconductors

Amorphous materials could be described as a material in which atom structures repeat only to a finite extent. They do not have a long-range (more than 10 nm) periodicity in its structure. Widely used amorphous semiconductors include chalcogenides, hydrogenated amorphous silicon, and oxides such as metal oxides or transparent conductive oxides. (Kugler & Shimakawa, 2015) An example of how the semiconductor might look like has already been displayed in Figure 5. It must also be noticed that, as in the case of crystalline materials, different fabrication conditions would result in different structures. The degree of disorder of the structure would change, as well as the amount of defects in the material.

In general, due to the amorphous structure, more defects such as voids can be found in these materials. In addition, other possible defects that appear are dangling bonds. They are possible bonds within an atom that have not been filled, depending on the valence electrons of the relevant atom. A representation of this phenomenon is outlined in Figure 6. These defects can be thermodynamically described, considering the disordered nature of the lattice, aiding the semiconductor properties description and interpretation. (Bar-Yam & Joannopoulos, 1987) These bonds are especially damaging to the performance of the semiconductor device because they create energy states in the middle of the bandgap, highly facilitating the occurrence of SRH recombination. It must be noted that the impact of these dangling bonds on the performance of a semiconductor device can be minimized by what it is called “passivating” the dangling bonds. In that instance, small atoms such as hydrogen are placed in the dangling bonds, allowing for that extra electron to be attached to another atom and form a covalent bond. (Lemaire & Gaspard, 1981)

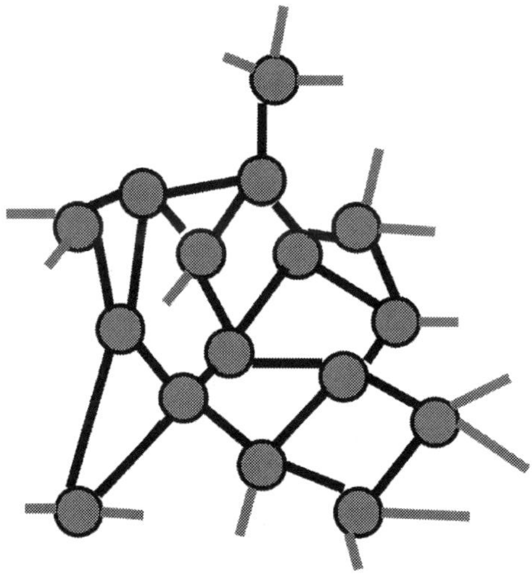

Figure 6. Representation of dangling bonds (red) in silicon amorphous structure

Another factor to take into account with amorphous materials is the fact that they might degrade upon illumination, according to the so-called Staebler-Wronski effect. (Staebler & Wronski, 1977) This effect was discovered in the case of amorphous silicon solar cells, but would also apply to other devices and materials. When an amorphous semiconductor is illuminated, charge carriers would be created. However, due to the high amount of defects, they would often recombine, liberating a certain amount of energy, which might affect the material, creating metastable defects. Although these defects can be reversed by applying an annealing process, the added complexity of the process is an additional barrier for amorphous materials. (Kolodziej, 2004)

Due to the abundance of defects in these kinds of structures, the conductive properties are not optimal for many applications. However, there are also some advantages of amorphous materials. As a general trend, amorphous materials tend to have a higher absorption coefficient. (Stuke, 1970) This implies that amorphous materials would be able to absorb the same light as their crystalline counterparts with a much thinner layer. This feature can be beneficial for a variety of applications, especially optical applications. (Yndurain, 1992) In addition, these materials can usually be fabricated at lower temperatures and higher deposition rates, making them potentially more cost-effective than their crystalline counterparts.

Polycrystalline and Nanocrystalline Semiconductors

Polycrystalline and nanocrystalline semiconductor materials display a structure in between complete order and complete disorder. In general, they consist of several crystals, which are in contact by grain boundaries. Depending on how big the crystals are, it would be classified either as polycrystalline (centimeters to millimeters) or nanocrystalline materials (micrometers to nanometers) (Trindade, 2001). An example of these kinds of materials is given in Figure 7. It must also be noticed from this image that, in general, nanocrystalline structures are often combined with amorphous material around it. The prime example of a material that can show both a nanocrystalline and a polycrystalline structure is silicon, although other materials such as GaAs or Cadmium Indium Gallium Selenide (CIGS) can also present these kinds of structures.

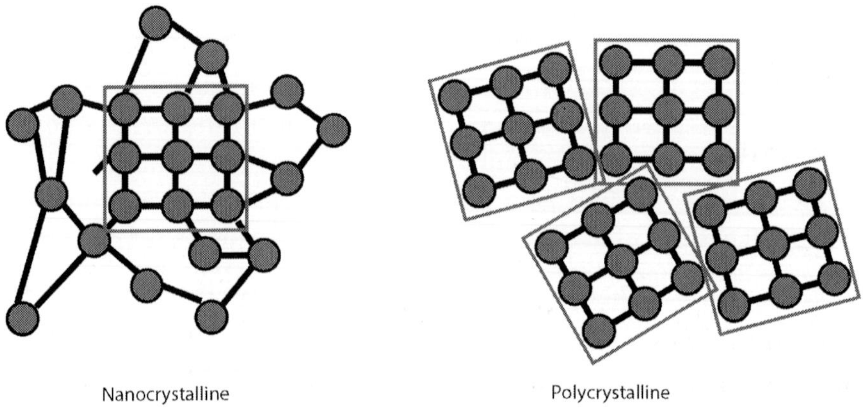

Figure7. Schematic nanocrystalline and polycrystalline structures

These kind of structures add a new defect to the previously described ones, the grain boundaries. The grain boundaries are described as the contact points between two monocrystals. This generally causes dangling bonds, which can create recombination centers and would decrease the efficiencies of the devices. Passivation by hydrogen introduction helps the passivation of these bonds, similarly to the case of amorphous materials. (Kurobe, et al., 2001)

It must also be taken into account that the shape of the crystal also has an effect on the performance of the devices. Long thin crystals in one direction would enhance the conductivity in that direction, while suppressing it in other directions, making the properties highly anisotropic.

On the other hand, if the crystals are more homogeneous in size, all the electric current produced will be equally likely to move in any direction, resulting in a more isotropic material. (Kurobe, et al., 2001) Depending on the application of the material, one of these instances will be preferred. The growth of these crystals can then be controlled by the growth method and conditions, resulting in the desired material. However, it must be noted that these materials are very sensitive to the deposition conditions, and thus, methods that can be precisely controlled are needed.

Nevertheless, it must be taken into account that when nanocrystals are included in a material matrix, the electrical properties would be affected. A great amount of grain boundaries are introduced in the material, and thus the recombination rates would be increased significantly. Thus, some passivation steps must be introduced in order to mitigate the effects of grain boundaries. (Alivisatos, 1996) In addition, the deposition methods for these materials at the moment are usually rather time consuming, reducing the industrial throughput and cost effectiveness. That is an issue that should also be addressed for large-scale implementation. (Kovalenko, 2015)

PHOTOELECTROCHEMICAL MATERIALS

So far, the discussion has been centered in the semiconductor materials and their properties in general. However, PEC devices have their own requirements regarding the properties of the chosen semiconductors, as shown in the previous chapter. An ideal semiconductor photoelectrode must absorb light efficiently, separate and transport charge carriers, and be able to drive the HER and OER at the surface, which is tied with the band alignment. Moreover, it must be stable in solution and be made out of earth-abundant materials to be cost effective. However, looking at the available semiconductors, no material so far fulfills all of these challenges and demands, and thus trade-offs need to be established. In addition, some material combinations can be designed to come closer to tackle these different issues, and different strategies can be implemented. (Krol & Gratzel, 2012)

The most common challenges that must be kept on mind are the inherent poor stability and the general high costs that are tied to good performances, such as the ones used for GaAs and $GaInP_2$. The other main option would be to use metal oxides, which are reasonably stable and cheap. However, one of the challenges of this material is the performance, with modest light absorption and challenged charge carrier transport and collection due to the numerous defects in the structures. In addition, due to their band alignment, materials can often only be used either as photoanode, to drive the OER reaction, or as photocathode, preferring the HER reaction. Thus, semiconductor materials are usually divided in two categories: those that can be uses as photoanodes and those applied to photocathodes.

The specific materials that can be used as photoanode and photocathode are discussed in this chapter. In addition, more details will be given in this section regarding the performances and costs of each material and the possible strategies that can improve these efficiencies.

Photoanode Materials

Photoanodes are the ones able to drive the OER, which is assumed to be the most challenging of the two half-reactions since according to the half reaction stoichiometry, they involve four electrons instead of two. Transition metal oxides are often seen as valid candidates for photoanodes, since they fulfill several of the outlined requirements, namely good band alignment, reasonable bandgap and stability. These oxides include metals such as Ce, Ti, Zr, Nb, Ta, Zn, Ga, Ge, Sn or Sb. Additionally, group 1, 2 and 3 metal ions can be added to help the growth of a certain crystal structure. (Chen, et al., 2013)

In particular, TiO_2 has been extensively studied, since the first experiments by Fujishima and Honda. (Fujishima & Honda, 1972) This material is naturally an n-type material due to the oxygen in the structure, and thus it is an effective photoanode for OER. The band alignment is favorable and although its bandgap is slightly high it can still absorb a certain amount of the solar spectrum. The performance achieved by (Fujishima & Honda, 1972) is shown in Figure 8, achieving current densities of more than 1 mA/cm^2. Moreover, it is also a relatively stable material to conduct the water splitting reaction and a rather cheap alternative, making it a suitable material for solar water splitting. (Hu, et al., 2014)

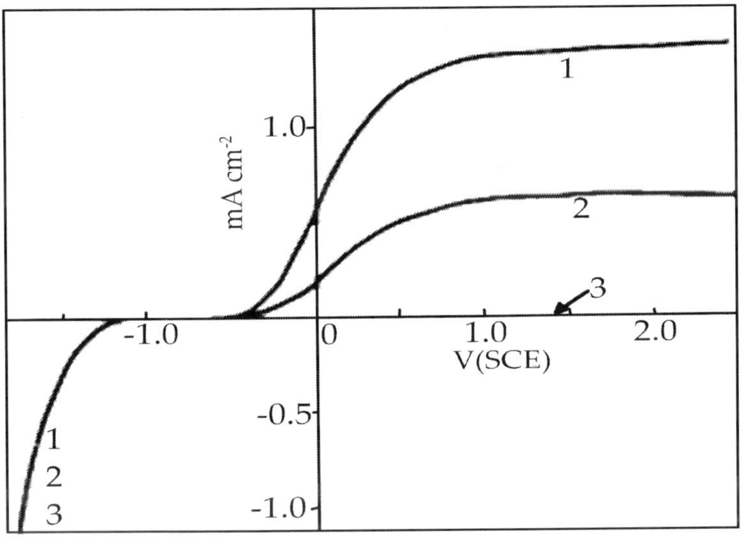

Figure 8. Current density vs. voltage characteristics for a TiO_2 n-type semiconductor. The TiO_2 is a single crystal of rutile materials used after annealing at 700°C and vacuum for roughly 4 h to increase the conductivity of the crystal. The wafer was approximately 1.5 mm thick with an area of 1 cm^2. Indium and a copper wire were used as an ohmic contact. From (Fujishima & Honda, 1972)

However, the potential of the conduction band (CB) is only slightly positive, so that the HER does not easily happen unless the solution pH is changed or the semiconductor is doped with other compounds such as Sr or Ba, which will give it a perovskite structure. (Lee, et al., 2010) Other possible perovskite structures are $SrTiO_3$ and $NaTaO_3$. However, the 4.0 eV bandgaps., although able to drive the reaction, is too large for efficient light absorption to create a reasonable amount of charge carriers. The inclusion of more highly charged counter ions into titanates, niobates and tantalates ($La_2Ti_2O_7$, $K_2La_2Ti_3O_{10}$, $Ba_5Nb_4O_{15}$, $Sr_2Ta_2O_7$ and $Ba_5Ta_4O_{15}$) can further increase the performance. In addition, it is important to know that d-10 configuration oxides have in general a lower performance due to the high electronegativity. (Chen, et al., 2013)

Another downside of using TiO_2 as the absorber material is the relatively low absorption of solar light, steaming from the high bandgap of the material. Strategies to tackle this issue have also been proposed, such as the inclusion of carbon doping (Park, et al., 2006) plasmonic structures (Zhang, et al., 2013) or dyes. (Oskam, et al., 2001) However, these approaches are either rather complex or expensive, or they introduce instabilities and toxic materials in the system. Thus other more absorbent materials must be explored.

Regarding the semiconductor absorption, metal oxides with strong covalent bonds tend to have a stronger visible light absorption. In particular, Mo and W compounds and oxide tend to have a positive response to visible light. In particular, WO_3 photoanodes, with a bandgap off 2.7 eV, are able to efficiently produce O_2 from water oxidation under solar light. In addition, other oxides or compounds based on Cu, Cd, In, Tl, Sn , Pb or Bi are either yellow or red, indicating light absorption, and allowing for solar water splitting. In particular, monoclinic $BiVO_4$ has a bandgap of 2.4 eV and an adequately VB alignment with the OER. Another very popular oxide that is often used as a photoanode is hematite (Fe_2O_3). This is an abundant, stable material with an adequate bandgap for solar light absorption and high refractive index, absorbing light up to the infrared region. However, the charge carrier lifetime inside the material is rather low due to the high concentration of defects,. (Sivula, 2012) In addition, it was found that these properties suffered from high anisotropies, and thus the crystal orientation within the material would also play an important role in the improvement of this material. (Goodenough, 1971) Nanostructures (Sivula, et al., 2010) and doping strategies (Ling, et al., 2011) have been successfully applied to improve the performance of hematite photoanodes, achieving current densities of more than 2 mA/cm^2 at a voltage of 1.23 V vs. RHE. Some results regarding the doping of hematite by different elements is summarized in

Table 2: PEC performances of different hematite photoanodes with different element doping. From (Li, et al., 2013)

Dopant	Concentration	Technique	Photoresponses	Reference
W	0.0025 %	Synthesized ceramics	IPCE 29 % at 400 nm and 1.23 V vs RHE	(Shinar & Kennedy, 1982)
Mo	15 %	Electrodepo-sition	IPCE 12 % at 400 nm and 1.4 V vs RHE	(Kleiman-Shwarsctein, et al., 2008)
V	0.05 %	Synthesized ceramics	IPCE 28 % at 400 nm and 1.23 V vs RHE	(Shinar & Kennedy, 1982)
Nb	0.5 %	Synthesized ceramics	IPCE 27 % at 400 nm and 1.23 V vs RHE	(Shinar & Kennedy, 1982)

Nb	0.1 %	Synthesized ceramics	135 μA/cm^2 at 0.2 V vs. SCE (100 W Hg lamp, 20 mW/cm^2)	(Houlihan, et al., 1985)
Nb	1.5 %	Single crystal	IPCE 37 % at 370 nm and 1.23 V vs RHE	(Sanchez, et al., 1988)
Nb	10 %	Synthesized ceramics	IPCE 26.7 % at 450 nm and 1.5 V vs RHE	(Aroutiouniana, et al., 2006)
Ta	0.5 %	Synthesized ceramics	IPCE 32 % at 400 nm and 1.23 V vs RHE	(Shinar & Kennedy, 1982)
Ta	0.5 %	Synthesized ceramics	640 μA/cm^2 at 0.7 V (two electrode setup, high pressure lamp of 1 kW)	(Aroutiounian, et al., 2002)
Si	2 %	Synthesized ceramics	IPCE 34 % at 400 nm and 1.23 V vs RHE	(Shinar & Kennedy, 1982)
Si	2 %	Synthesized ceramics	IPCE 9.3 % at 340 nm and 1.49 V vs RHE	(Kennedy, et al., 1981)
Si	1.5 %	APCVD	IPCE 42 % at 370 nm and 1.23 V vs RHE	(Kay, et al., 2006)
Si	5.3 %	RMS	IPCE 2.5 % at 375 nm and 1.54 V vs RHE	(Glasscock, et al., 2007)
Si	0.1 %	SP	IPCE 23 % at 360 nm and 1.4 V vs RHE	(Liang, et al., 2008)
Si	0.2 %	SP	0.37 mA/cm^2 at 1.23 V vs RHE (0.8 sun, AM1.5)	(Liang, et al., 2008)
Ti	5 %	RMS	IPCE 14 % at 375 nm and 1.54 V vs RHE	(Glasscock, et al., 2007)
Ti	4 %	RBD	IPCE 31 % at 360 nm and 1.4 V vs RHE	(Hahn & Mullins, 2010)
Ti	2.5 %	PVD	IPCE 14 % at 375 nm and 1.54 V vs RHE	(Kronawitter, et al., 2011)
Ti	1 %	Synthesized ceramics	IPCE 1.2 % at 340 nm and 1.49 V vs RHE	(Kennedy, et al., 1981)
Ge	0.05 %	Synthesized ceramics	IPCE 5.5 % at 340 nm and 1.44 V vs RHE	(Kennedy, et al., 1981)
Zr	0.5 %	Synthesized ceramics	IPCE 23 % at 400 nm and 1.23 V vs RHE	(Shinar & Kennedy, 1982)
Zr	2 %	Electrodeposition	2.1 mA/cm^2 at 1.64 V vs RHE (Xe lamp 150 mW/cm^2)	(Kumar, et al., 2011)
Ru	5 %	Synthesized ceramics	108 μA/cm^2 at 0.2 V vs SCE (100 W Hg arc, 20 mw/cm^2)	(Houlihan, et al., 1985)

Sn	4 %	RBD	IPCE 21 % at 360 nm and 1.4 V vs RHE	(Hahn & Mullins, 2010)
Sn	0.05 %	Synthesized ceramics	IPCE 4 % at 340 nm and 1.04 V vs RHE	(Kennedy, et al., 1981)
Sn	4 %	Deposition annealing	IPCE 0.7 % at 420 nm and 0.3 V vs Ag/AgCl	(Jang, et al., 2009)
Sn	9.4 %	Hydrothermal method	IPCE 19.2 % at 400 nm and 1.23 V vs RHE and 1.86 mA/cm^2 at 1.23 V vs. RHE (AM1.5)	(Ling, et al., 2011)
Ce	0.05 %	Synthesized ceramics	IPCE 30 % at 400 nm and 1.23 V vs RHE	(Shinar & Kennedy, 1982)
Hf	0.002 %	Synthesized ceramics	IPCE 26 % at 400 nm and 1.23 V vs RHE	(Shinar & Kennedy, 1982)
Pt	5 %	Electrodepo-sition	IPCE 12 % at 400 nm and 1.4 V vs RHE	(Hu, et al., 2008)
Pt	5 %	Hydrothermal	1.38 mA/cm^2 at 1.25 V vs RHE under 1 sun illumination	(Cha, et al., 2010)
Al	0.46 %	Electrodepo-sition	IPCE 8 % at 400 nm and 1.4 V vs RHE	(Kleiman-Shwarsctein, et al., 2008)
Cr	5 %	Electrodepo-sition	IPCE 6 % at 400 nm and 1.4 V vs RHE	(Kleiman-Shwarsctein, et al., 2008)
In	8-20 %	Sputtering	EQE 21 % at 400 nm and 1.5 V vs RHE	(Schumacher, et al., 1990)
Mg	1 %	Plasma sprayed	2.1 % at 400 nm and -0.4 V vs SCE (pH 13) cathodic 0.84 % at 400 nm and +0.5 V vs SCE (pH 13) anodic	(Parent, et al., 1987)
Ca	1 %	Synthesis ceramics	-	(Gori, et al., 1980)
Ni	5 %	Spray pyroly-sis	0.26 mA/cm^2 at 0.59 V vs. RHE (500 W Xe lamp)	(Sartoretti, et al., 2005)
Cu	5 %	Spray pyroly-sis	66.8 μA/cm^2 at 1.61 V vs RHE (150 W Xe lamp)	(Satsangi, et al., 2008)
Zn	5 %	Spray pyroly-sis	422.6 μA/cm^2 at 1.61 V vs RHE (150 W Xe lamp)	(Satsangi, et al., 2008)
Zn	5 %	Spray pyroly-sis	0.64 mA/cm^2 at 1.71 V vs RHE (150 W Xe lamp)	(Kumari, et al., 2006)

Cd	1 %	Electrodeposition	1.25 mA/cm^2 at 1.23 V vs RHE (AM1.5)	(Bak, et al., 2011)
Si + Ti	4 % + 2 %	USP	IPCE 37 % at 370 nm and 1.23 V vs RHE	(Zhang, et al., 2010)
Be + Sn	6 % + 4 %	Deposition annealing	IPCE 1.2 % at 420 nm and 0.3 V vs Ag/AgCl (0.2 M NaOH)	(Jang, et al., 2009)
Ti + Al	5 % + 1 %	Spray pyrolysis	IPCE 25 % at 400 nm and 0.7 V vs RHE	(Sartoretti, et al., 2005)
Ti + Li	5 % + 1%	Spray pyrolysis	0.9 mA/cm^2 at 0.5 V vs RHE (500 W Xe lamp)	(Sartoretti, et al., 2005)

However, as already mention, charge separation and collection in these materials are still challenging due to the disordered material structure. (Chen, et al., 2013) Thus, other more ordered materials must be explored. It must be noticed that most high performance photoanodes based on Ti, Nb or Ta based on a perovskite structure. (Walter, et al., 2010) (Aroutiounian, et al., 2005) In particular, $SrNbO_2N$ has been proven a viable alternative to more simple materials, with a bandgap of 1.8 eV and achieving near stoichiometric HER and OER. (Maeda, et al., 2011) Nevertheless, these materials, although stable and able to absorb a high portion of the sun radiation, still have relatively low photocurrents, and moreover, the materials needed for their synthesis are not as abundant as the elements previously commented such as titanium dioxide or hematite.

Doping the material in a way that it can maintain the crystal lattice produced additional free charge carriers, dramatically increasing the electrical conductivity, and thus the overall performance of the photoanode. An example is Zn doped Ga_2O_3. (Walter, et al., 2010) Oxygen vacancies can also be created as an n-type dopant, and can be produced by higher temperatures and lower O_2 concentrations. Doping opens a wide range of materials that could be further studied and used as photoanodes. Some sources argue that there are more than 130 possible candidate materials, (Walter, et al., 2010) and thus it is not possible to discuss them all within the scope of this book. However, the most successful and studied materials have been summarized in Table 4. Some of the most interesting ones are TiO_2 and ZnO, which have similar bandgap and band edge position. However, the ZnO performance is significantly lower. In addition, Fe_2O_3, WO_3, $SrTiO_3$, $KTaO_3$ and $BaTiO_3$ have been investigated intensively, but with still low performances due to their large bandgaps. That is why Fe_2O_3 and WO_3 are the most studied materials, since they have the narrower bandgaps and better stabilities. (Li, et al., 2013) Regarding non-oxides, Si, InP, GaP, $GaInP_2$, CdS, CdSe, CdTe or a-SiC:H have been studied. Unfortunately, most of them suffer from corrosion under light irradiation. One of the most promising materials is highly purified planar crystalline Si, which can be a photocathode or a photoanode depending on the doping. (Li, et al., 2013) Looking at some more complicated compounds, $BiVO_4$, Biu_2VO_6, TaON, Ta_3N_5, $CuGa_3Se_5$ and $CuWO_4$ are among the most studied ones, with Bi and Ta compounds as the most promising ones, especially in terms of bandgap and light absorption. (Li, et al., 2013)

Table 3: PEC responses of the most successful photoanodes based on earth-abundant materials. Adapted from (Li, et al., 2013)

Material	Preparation	Catalyst	Electrolyte	Photoresponse	Reference
$BiV_0$4	MOD	-	0.5 M Na_2SO_4	IPCE 33 % at 400 nm and 29 % at 420 nm (1.9 V vs RHE)	(Sayama, et al., 2003)
$BiV_0$4	MOD	-	0.5 M Na_2SO_4	IPCE 20 % at 420 nm and 1.6 V vs. RHE	(Sayama, et al., 2006)
$BiVO_4$	MOD	Ag ion treat-ment	0.5 M Na_2SO_4	IPCE 44 % at 420 nm and 1.6 V vs. RHE	(Sayama, et al., 2006)
$BiV_0$4	Powder spreading	-	0.5 M Na_2SO_4	IPCE 1.5 % at 420 nm and 1.6 V vs. RHE	(Long, et al., 2008)
$BiVO_4$	Powder spreading	Co_3O_4	0.5 M Na_2SO_4	IPCE 6.5 % at 420 nm and 1.6 V vs. RHE	(Long, et al., 2008)
$BiVO_4$	MOD	-	0.5 M Na_2SO_4	0.173 μA at 1.1 V vs. RHE	(Luo, et al., 2008)
$BiVO_4$	MOD	-	Saturated $NaHCO_2$ with CO_2 gas bub-bling	IPCE 32 % at 420 nm and 1.23 V vs. RHE	(Sayama, et al., 2010)
$BiVO_4$	Powder spreading	-	0.5 M K_2SO_4	IPCE 12 % at 440 nm and 1.5 V vs. Ag/AgCl	(Iwase & Kudo, 2010)
$BiVO_4$/RGO	Powder spreading	-	0.5 M Na_2SO_4	IPCE 4.2 % at 400 nm and 0.75 V vs Ag/AgCl	(Ng, et al., 2010)
$BiV_0$4	CBD	-	0.5 M Na_2SO_4	1.4 mA/cm^2 at 1.6 V vs. RHE	(Luo, et al., 2010)

$BiVO_4$	CBD	-	0.5 M Na_2SO_4	0.4 mA/cm^2 under AM1.5 (1.6 V vs. RHE)	(Feng, et al., 2011)
$BiVO_4$	USP	-	0.5 M Na_2SO_4	IPCE 1 % at 420 nm and 0.3 V vs SCE	(Jing, et al., 2010)
1 % W - $BiVO_4$	USP	-	0.5 M Na_2SO_4	IPCE 10 % at 400-450 nm and 0.3 V vs SCE	(Jing, et al., 2010)
$BiVO_4$	Reactive ballistic deposition	-	0.5 M Na_2SO_4	IPCE over 21 % at 340-460 nm	(Berglund, et al., 2011)
7 % W – $BiVO_4$	MOD	CoPi	0.1 M K_3PO_4 buffered to pH 8	IPCE 33 % at 370 nm and 1.23 V vs RHE	(Zhong, et al., 2011)
BiV_O4	Spray pyrolysis	-	0.15 M K_2SO_4 (pH 7)	IPCE 6 % at 450 nm and 1.63 V vs. RHE	(Liang, et al., 2011)
FTO/SnO_2/$_B$iVO4	Spray pyrolysis	-	0.15 M K_2SO_4 (pH 7)	IPCE 25 % at 450 nm and 1.63 V vs. RHE	(Liang, et al., 2011)
FTO/SnO_2/1 % W - $BiVO_4$	Spray pyrolysis	-	0.15 M K_2SO_4 (pH 7)	IPCE 46 % at 450 nm and 1.63 V vs. RHE	(Liang, et al., 2011)
3 % Mo-BiV_O4	MOD	RhO_2	0.5 M K_2SO_4	3.3 mA/cm^2 under AM 1.5G (1.6 V vs. RHE)	(Luo, et al., 2011)
3 % Mo-$BiVO_4$	MOD	-	Seawater	IPCE 41 % at 420 nm and 1 V vs. RHE	(Luo, et al., 2011)
3 % Mo-$BiVO_4$	MOD	RhO_2	Seawater	IPCE 49 % at 420 nm and 1 V vs. RHE	(Luo, et al., 2011)
W-$BiVO_4$ (Bi/V/W = 4.5/5/0.5)	Deposition - annealing	-	0.1 M Na_2SO_4	IPCE 58 % at 400 nm and 0.3 V vs. Ag/AgCl	(Ye, et al., 2010)
W-$BiVO_4$ (Bi/V/W/Mo = 4.6/4.6/0.2/0.6)	Deposition - annealing	-	0.1 M Na_2SO_4	IPCE 8 % at 400 nm and 0.6 V vs. Ag/AgCl	(Park, et al., 2011)
2 % Mo – $BiVO_4$	MOD	CoPi	0.5 M $NaSO_4$ (phosphate buffer at pH 7)	IPCE 35 % at 450 nm and 0.9 V vs. ag/AgCl	(Pilli, et al., 2011)
FTO/SnO_2/$BiVO_4$	Spray pyrolysis	CoPi	0.15 M K_2SO_4 buffered to pH 5.6 with 0.09 M $K_2H_2PO_4$/0.01 M K_2HPO_4	1.7 mA/cm^2 under AM1.5G and 1.23 V vs. RHE	(Abdi & Krol, 2012)
WO_3/BiV_O4	MOD	-	0.5 M Na_2SO_4	IPCE 23 % at 420 nm and 1 V vs Ag/AgCl	(Chatchai, et al., 2009)

$WO_3/BiVO_4$	Deposition-annealing	-	0.5 M Na_2SO_4	IPCE 38 % at 420 nm and 0.7 V vs Ag/AgCl	(Hong, et al., 2011)
$WO_3/BiVO_4$	CBD	-	0.5 M Na_2SO_4	IPCE 31 % at 420 nm and a bias of 0.5 (two electrode setup) 1.6 mA/cm^2 at 1 V bias back-side AM 1.5G	(Su, et al., 2011)
$BiVO_4$	Electro-deposition	FeOOH	0.1 M KH_2PO_4	IPCE 56 % at 410 nm and 1.2 V vs RHE	(Seabold & Choi, 2012)
PO_4 doped BiV_O4	EPD	-	0.5 M Na_2SO_4	0.548 ma/cm^2 at 1.23 V vs RHE (300 W Xe lamp)	(Jo, et al., 2012)
$BiVO_4/WO_3$	MOD	-	0.1 M $KHCO_3$ with CO_2 bubbling	IPCE 31 % at 420 nm and 1.2 v vs RHE	(Saito, et al., 2012)
$BiVO_4/SnO_2/WO_3$	MOD	-	0.1 M $KHCO_3$ with CO_2 bubbling	IPCE 43 % at 420 nm and 1.2 V vs RHE and 2.82 mA/cm^2 at 1.23 V vs RHE (AM 1.5)	(Saito, et al., 2012)
$BiVO_4$	EPD	CoPi, CoO_x, IrO_x, MnO_x	0.1 M $KaPO_4$ containing 0.02 M $NaIO_3$ pH 7	Photocurrent density CoPi/$BiVO_4$ > $CoO_x/BiVO_4$ > $IrO_x/BiVO_4$ > MnO_x/B_iVO4	(Wang, et al., 2012)
FTO/$SnO_2/BiVO_4$	MOD	-	0.5 M Na_2SO_4	IPCE 6.2 % at 350 nm and 1.6 V vs. RHE	(Chatchai, et al., 2008)
Ta_3N_5	Reactive sputtering	IrO2	0.1 M Na_2SO_4 pH 8.5	2.3 mA/cm^2 under irradiance of 300 W Xe lamp	(Yokoyama, et al., 2011)
Ta_3N_5	Oxidation and nitration of Ta foil	-	0.1 M K_2SO_4 adding KOH pH 11	1 % at 420-550 nm and -0.1 V vs Ag/AgCl (0.75 V vs RHE)	(Ishikawa, et al., 2004)
TaON	EPD	IrO_2	0.1 M Na_2SO_4 pH 6	76 % at 400 nm and 0.6 V vs. Ag/AgCl	(Abe, et al., 2010)
Ta_3N_5	EPD	IrO_2	0.1 M Na_2SO_4 pH 6	31 % at 500 nm and 1.15 V vs RHE	(Higashi, et al., 2011)
Ta_3N5	EPD	Co_2O_4	1 M NaOH	26 % at 400 nm and 24 % at 550 nm at 1.2 V vs RHE 40 % at 400 nm and 36 % at 550 nm at 1.4 V vs RHE	(Liao, et al., 2012)

TaON	Anodization and nitration of Ta foil	-	1 M KOH	2.6 mA/cm^2 at 1.52 V vs RHE (AM 1.5)	(Banerjee, et al., 2009)
Ta_3N5	Anodization and nitration of Ta foil	-	1 M KOH	5.3 % at 450 nm and 0.5 V vs Pt pseudo-reference	(Feng, et al., 2010)
TaON	TaON powder spreading	-	0.1 M $_{N}a_2$SO4	0.023 mA/cm^2 at 1 v vs. Pt pseudo reference under 300 W Xe lamp in two electrode configuration	(Abe, et al., 2010)
$_{T}a_3$N5	$Ta_3N_{5\ powder}$ spraying	-	0.1 M Na_2SO_4	0.10 mA/cm^2 at 1.55 v vs RHE (300 W Xe lamp)	(Maeda & Domen, 2009)
$Ta_xCo_{1-x}N_y$	Dip coating and nitration	Co oxide	0.1 M Na_2SO_4 adding NaOH pH 11	25 % at 450 nm and 1.45 V vs. RHE	(Cong, et al., 2012)
TaON	EPD	CoO_x	Sodium phosphate buffer solution, pH 8	42 % at 400 nm and 1.45 V vs. RHE	(Higashi, et al., 2012)
Ta_3N_5	Anodization and nitration of Ta foil	IrO_x	0.1 M Na_2SO_4 adding KOH pH 11	IPCE 10 % at 400 nm and 1.45 V vs. RHE	(Cong, et al., 2012)
Ta_3N_5	Evaporation of Ta metal followed by nitration	Co^{2+}	1 M KOH, pH 13.6	IPCE 9 % at 500 nm and 1.2 V vs. RHE	(Dang, et al., 2012)
WO_3	CBD	-	1 M $HClO_4$	IPCE 75 % at 400 nm and 0.5 V vs. RHE and 2.2 mA/cm^2 at 1.23 V vs RHE and 1 sun	(Ulmann & Augustynski, 2001)
n-CdTe/NiO_x	Sputtering	-	1 M KOH (pH 14)	22.9 mA/cm^2 at 1.15 V vs RHE	97
np^+-Si/NiO_x	Sputtering	-	1 M KOH (pH 14)	30 mA/cm^2 at 1 V vs RHE	98
ZnO/Ta_2O_5	ALD	-	0.1 M KOH (pH 13)	0.8 mA/cm^2 at 0.5 V vs RHE	145

From this analysis, it can be seen that several materials have recently gotten more attention in the recent years. Some of them, such as $BiVO_4$ and (oxy) nitrides such as TaON and Ta_3N_5, is worth taking a closer look.

Monoclinic $BiVO_4$

This material has attracted a lot of interest for water oxidation under visible illumination, mainly due to its attractive bandgap of 2.4 eV and band alignment, and its non-toxic nature. $BiVO_4$ exists in three main crystalline structures: monoclinic scheelite, tetragonal and octagonal. In principle, monoclinic scheelite is the most active and effective form, being able to achieve a theoretical STH efficiency limit of 9.1 %. (Sayama, et al., 2006) first reported a photoelectrode based on $BiVO_4$ with an IPCE around 30 % at 1.9 V. Since then, several advances have been made in this material, including doping strategies and nanostructuring. (Abdi, et al., 2013) In particular proposed a gradient doping to increase charge-separation, which greatly improved the photoanode performance.

Another factor to take into account is the morphological details of the photoanode, such as size, shape and connectivity of the particles. For example, the crystals might be more reactive at a certain orientation or facet, which can greatly affect the photoconversion efficiency since they affect the thermodynamic and kinetic factors. Moreover, nanostructures can also help with the challenge of achieving the desired crystal orientation and plane exposure, which would be mainly controlled by the deposition methods. (Gan, et al., 2014)

Finally, the issue of choosing the right catalyst must be considered. Some studies indicate that high STH efficiencies can already be achieved by using a cobalt based catalyst, also called CoPi, electrodeposited at the surface. This is especially encouraging, since this would mean that a photoanode based on completely earth-abundant materials can be achieved. (Wang, et al., 2012)

However, there are still some challenges to this photoanode material. First, voltages higher than 1 V are needed to achieve reasonable current densities. This fact might be related the high surface recombination at the semiconductor/electrolyte interface. The deposition techniques have a big role in this issue, where more straight-forward methods like drop casting achieve lower efficiencies than the time-consuming spray pyrolysis methods. In addition, there are some studies that reported certain instabilities of the material (Sayama, et al., 2006), although more recent studies reported relatively high stability. (Abdi, et al., 2013)

TaON and Ta_3N_5

These materials combine a narrow bandgap (2-2.2 eV), which allows solar light absorption, with good band alignment for H_2 production, with the valence and the conduction bands levels corresponding with the oxidation and reduction potentials, respectively. In particular, Ta_3N_5 has been given much attention in the past few years because of its ability to collect more than 45 % of the solar energy and a band alignment that allows unassisted water splitting. Moreover, the theoretical STH efficiencies, based in the bandgap and light absorption, is up to 15.9 %. (Gan, et al., 2014)

Despite these favorable characteristics, the efficiencies achieved with the first demonstrators are still low, possibly due to the insufficient crystallization of the precursor materials, which enhances charge carrier recombination. Moreover, it suffers from poor conductivity and stability. (Gan, et al., 2014) Fortunately, (Higashi, et al., 2011) made substantial contributions to the field by fabricating an efficient PEC device based on these materials, highlighting the importance of efficient contacts among the material particles to promote charge carrier transport within the material. In addition, they determined that a key factor on achieving a good particle contact lies within the fabrication method and the parameters used, since good contact cannot be achieved by a simple temperature treatment. Alternatively, they used a so-called post-necking treatment under NH_3 atmosphere, in which bridges are formed between the different particles, allowing for charge carriers pathways.

Another way to improve the conductivities and charge carrier densities of these materials is to include appropriate ion doping. Metal alloys and salts can be used for this purpose, such as the cobalt phosphate doping proposed by (Li, et al., 2013). These photoanodes had better crystallinity and smaller particles with smoother surfaces, demonstrating an improvement in photocatalytic activity under visible light. In addition, heterojunctions alternating TaON and Ta_3N_5 can improve charge carrier separation, highly improving the performance of these photoanodes. Such structures exhibit a photocurrent density of 1.2 mA/cm^2 under visible light at 0.5 V vs. SCE, an excellent photocatalytic activity attributed to the change in energy levels caused by the heterojunction itself. Moreover, this device also exhibited a high stability, due to the strong bonds between the two composites. However, it must be taken into account that another instability occurs in these materials, since TaON tends to suffer from self-oxidation, which deactivates the catalytic activity. This could be solved by some extent by including highly selective catalysts that favor water oxidation over TaON oxidation, such as IrO_x. However, this also includes additional costly elements into the structure. (Gan, et al., 2014)

To further improve these photoanodes, there are several strategies. First, the band structure can be tuned further by introducing other doping elements. In the case of a wide bandgap semiconductor like TiO_2, doping can lower the bandgap, allowing for more light absorption. However, doping also brings defects and may lower the charge carrier separation and collection and increase recombination, and thus it is better to utilize a semiconductor with an intrinsically narrow bandgap. These more absorbing materials can also be doped to increase the charge carrier concentration and diffusion length. Second, surface treatments can be included to improve the efficiency by removing segregation and defects at the surface. These phenomena occur naturally, and their density depends on the fabrication method. These states can be removed to some extent by etching the surface. Third, catalysts can be added to the surface, which could really lower the overpotentails, and thus increase the overall PEC performance. Metals such as Pt, Ru and Pd are the most common candidates, although their price and abundance is a limiting factor for their commercial use. Beside noble-metal, promising candidates are MoS_2, W, Ni, P, Cu, Fe or a combination of them. Fourth, the morphology of the material can be manipulated to increase the surface exposed to the solution while still maintaining the available light absorption. That can also reduce the diffusion length from the surface where the reaction occurs to the counter-electrode. Finally, including a multilayer design can improve the charge collection and injection at the surface, as well as passivate the surface and reduce electron leaking. (Li, et al., 2013)

III-V Materials

Another option to improve the performance of photoanodes is to use non earth-abundant materials, such as III-V materials. The most common materials in this category are GaP, GaAs or GaAlAs. Table 4 summarizes some of these results, which tend to be much higher than those reported by metal oxides and other earth-abundant materials. However, due to the scarcity and complex material processing of III-V materials, these devices would be very expensive, and thus unlikely to be used in a real device.

Table 4: Summary of PEC photoanodes based on III-V technologies. Adapted from (Hu, et al., 2015)

Material	Preparation	Catalyst	Electrolyte	Photoresponse	Reference
np^+-InP/NiO_x	Sputtering	-	1 M KOH (pH 14)	20.8 mA/cm^2 at 0.85 V vs RHE	(Sun, et al., 2015)
np^+-GaAs/NiO_x	ALD	Ni	1 M KOH (pH 13.7)	14 mA/cm^2 at -0.38 V vs SCE	(Hu, et al., 2014)
n-GaP	ALD	NiO_x	1 M KOH (pH 13.7)	2.5 at -0.16 V vs SCE	(Hu, et al., 2014)
n-GaAs, n-GaAlAs	Electrodeposition	-	Na_2SO_4 + NaOH	-0.6 mA/cm^2 at 1.5 V vs RHE	(Tomkiewicz & Woodall, 1977)

Photocathode Materials

Regarding the photocathode, not as much research has focused on these materials because the hydrogen half-reaction is not as challenging, since it only needs two electrons per hydrogen molecule. However, it is still worth researching, since these photoelectrodes seem to have their own issues to tackle . Several materials have been investigated for this purpose, including p-Si, p-InP, p-WSe_2, Cu_2O, p-GaP and CdS. The most efficient semiconductor would be InP combined with a Pt catalyst. However, these materials are very scarce, and thus not practical for practical applications. (Awad, et al., 2014) GaP can also be a good material for this application, but again the cost of the material is too high for practical applications. (Chen, et al., 2013)

A more affordable option would be to use metal oxides as photocathodes. However, despite many efforts in the last years, all the developed structures are far from the required performances. Most of them are due to the tendency to photocorrode. The ones that are more stable have high bandgaps, which have a limited light absorption and thus photocorrosion. (Awad, et al., 2014) Cu_2O is one of the most promising candidates, with photocurrent of 2 mA/cm^2, but with stability problems. It is interesting to notice that inclusion of Ti, Ni or Au improves the photostability. In addition, a layer of CuO on top of Cu_2O can also enhance the resistance to corrosion, also enhancing the current density at 0 V vs. RHE to almost 8 mA/cm^2. The drawback of these approaches is the increasing complexity and cost of the fabrication processes. Other option is to use doped materials, such as Mg doped Fe_2O_3, although the photocurrents are much lower. (Paracchino, et al., 2012)

Other materials that can be used as photocathodes are more complex oxides, such as $CuNbO_3$. This material has shown resistance to photocorrosion, although the photocurrents achieved are low, less than 1 mA/cm^2. $CuBi_2O_4$ has a better performance, with values around 5 mA/cm^2. (Berglund, et al., 2016) In addition, SiC can be a very efficient photocathodes if a p/i structure with a gradient doped p layer is introduced, showing that not only the material but also the device configuration is important. (Han, et al., 2015)

Table 5: Summary of the main results obtained for photocathodes based on earth-abundant materials for solar water splitting. Adapted from (Awad, et al., 2014)

Material	Preparation	Conditions	Photocurrent (mA/cm^2)	Reference
CuO	Solution based	AM1.5G, 1 M KOH and 0.55 V vs Ag/AgCl	-1.2	(Chiang, et al., 2012)
CuO	Sputtering	5 sun, 1M Na_2SO_4 at 0.2 V vs Ag/AgCl	-2.0	(Tang, et al., 2012)
$(Cu_{0.9}Ti_{0.1})O$	Sputtering	5 sun, 1M Na_2SO_4 at 0.2 V vs Ag/AgCl	-0.2	(Tang, et al., 2012)
Cu_2O	Electrodeposition	500 W Xe arc lamp, 0.1M Na_2SO_4 and 0 V vs Ag/AgCl	-0.5	(Mao, et al., 2012)
Cu_2O nanoparticles	Electrodeposition	500 W Xe arc lamp, 0.1M Na_2SO_4 and 0 V vs Ag/AgCl	-1.75	(Mao, et al., 2012)
Cu_2O	Polished plates and annealing	AM1.5G, 0.01 M KI + 0.0001 M I_2 and 0.3 V vs Ag/AgCl	-0.004	(Fernando, et al., 2001)
Cu_2O + Au, NiO	Polished plates and annealing	AM1.5G, 0.01 M KI + 0.0001 M I_2 and 0.3 V vs Ag/AgCl	-0.002	(Fernando, et al., 2001)
Cu-1.0/ Cu_2O/CuO	Electrodeposition	AM1.5G, 50 mM Na_2SO_4 and 0 V vs RHE	-1.54	(Zhang & Wang, 2012)
Cu_2O/ZnO/ Al_2O_3/ TiO_2	Electrodeposition	AM1.5G, 1 M Na_2SO_4 and 0 V vs RHE	-7.6	(Paracchino, et al., 2011)
α-Fe_2O_3	ALD	AM1.5G, 1 M KOH and 0.3 V vs RHE	-0.8	(Lin, et al., 2012)
pSi/ZnO	Sputtering	AM1.5G, 1 M KOH and 1.5 V vs Ag/AgCl	-1.5	(Sun, et al., 2012)
pSi/ZnO/ ZnO NW	Sputtering	AM1.5G, 1 M KOH and 1.5 V vs Ag/AgCl	-4	(Sun, et al., 2012)

NiO-P1 (dye)	Dropcast	Light emitting diode (>400 nm), phosphate buffer pH at 0.4 V vs Ag/AgCl	-0.225	(Li, et al., 2012)
NiO-P1 (dye) – Co1 (catalyst)	Dropcast	Light emitting diode (>400 nm), phosphate buffer pH at 0.4 V vs Ag/AgCl	-0.006	(Li, et al., 2012)
CuNb$_0$3	Solid-state synthesis	400 W Xe lamp, 0.5 M Na_2SO_4 and 0.6 V vs Ag/AgCl	-0.1	(Joshi, et al., 2011)
$CuNb_3O_8$	Solid-state synthesis	400 W Xe lamp, 0.5 M Na_2SO_4 and 0.6 V vs Ag/AgCl	-0.4	(Joshi, et al., 2011)
$CuBi_2O_4$	Electrodeposition	100 W Xe lamp, 0.1 M Na_2SO_4 and 0.5 V vs Ag/AgCl	-0.033	(Hahn, et al., 2012)
Cu$_5$Ta$_{110}$30	Solution based	400 W Xe lamp, 0.5 M Na_2SO_4 and 0.6 V vs Ag/AgCl and 250 °C	-1.5	(Fuoco, et al., 2012)
$Cu_5Ta_{11}O_{30}$	Solution based	400 W Xe lamp, 0.5 M Na_2SO_4 and 0.6 V vs Ag/AgCl and 350 °C	-2.6	(Fuoco, et al., 2012)
CaF$_{e2}$O4	Solution based	500 W Xe lamp, 0.25 M K_2SO_4 and 0.5 V vs Ag/AgCl	-0.002	(Matsumoto, et al., 1987)
$CaFe_2O_4$/Pt	Solution based	500 W Xe lamp, 0.25 M K_2SO_4 and 0.5 V vs Ag/AgCl	-0.250	(Matsumoto, et al., 1987)
La$_{2C}$uO4	Doctor blading	Thermo Oriel lamp KOH (PH 13) 0.4 V vs Ag/AgCl	-0.5	(Spijker, et al., 2008)
Ca doped La_2CuO_4	Doctor blading	Thermo Oriel lamp KOH (PH 13) 0.4 V vs Ag/AgCl	-2.5	(Spijker, et al., 2008)
Rh doped $SrTiO_3$	Pulse laser deposition	300 W Xe lamp, 0.1 M K_2SO_4 and 1 V vs Ag/AgCl	-0.8	(Kawasaki, et al., 2012)
$CuYO_2$	Solution based	200 W Xe lamp, 1 M KOH and 0.6 V vs Ag/AgCl	-0.027	(Trari, et al., 2006)
$CuFeO_2$	Electrodeposition	AM1.5G, 1 M NaOH and 0.4 V vs RHE	-0.7	(Read, et al., 2012)
CuO	Electrodeposition	AM1.5G, 0.5 M Na_2SO_4 and 0.6 V vs Ag/AgCl	-0.6	(Zheng, et al., 2012)
Cu-Ti-O/n-TiO_2	Anodization	AM1.5G, 0.1 M Na_2HPO_4 and 0 V vs Ag/AgCl	-0.25	(Mor, et al., 2008)
Cu_2O/TiO_2 heterojunction	Electrodeposition	Xe lamp, 0.1 sodium acetate and 1 V vs Ag/AgCl	-0.7	(Siripala, et al., 2003)
Cu_2O NW/bio-anode	Annealing Cu substrates	AM1.5G, 0.5 M Na_2SO_4 and 0 V vs Ag/AgCl	-0.05	(Qian, et al., 2010)

Dye coated NiO/$BiVO_4$	Solution based	300 W Xe lamp, 0.1 M Na_2SO_4 and 0 V vs Ag/AgCl	-0.0027	(Tong, et al., 2012)
CaFe$_{20}$4	Calcined at 1200 °C	500 W Xe lamp, 0.1 M NaOH and 0 V vs Ag/AgCl	-0.5	(Ida, et al., 2010)
$CaFe_2O_4$	Calcined at 1100 °C	500 W Xe lamp, 0.1 M NaOH and 0 V vs Ag/AgCl	-0.01 to -0.05	(Ida, et al., 2010)
a-SiC (p/i)/ Pt	PECVD	AM1.5G, 0.5 M KHP and 0 V vs RHE	-5.1	(Han, et al., 2015)
a-SiC (p/i)/ TiO_2/Ni-Mo	PECVD	AM1.5G, 0.5 M KHP and 0 V vs RHE	-8.3	(Digdaya, et al., 2015)

Similarly to the case of photoanode, the presented materials so far are based on earth-abundant materials, mainly metal oxides. Although these have achieved remarkable efficiencies, they are still rather low for an effective and economical commercial device. An alternative is to use III-V materials again, as presented in Table 6. Most of the materials used here are InP, or $GaInP_2$, since they are p-type materials. However, these materials are, as previously mention, rather expensive and that must be taken into account when a PEC practical device is designed, where probably light concentration must be considered.

Table 6. Summary of materials to be applied as photocathodes based on III-V materials. Adapted from (Hu, et al., 2015)

Material	**Deposition method**	**Cat-alyst**	**Electrolyte**	**Current density (mA/cm^2)**	**Reference**
p-InP	ALD	Pt	1 M $HCLO_4$ (pH 0)	24 mA/cm^2 at AM1.5	(Lin, et al., 2015)
p-InP nanotex-tured	ALD	Ru	1 M $HCLO_4$ (pH 0)	35 mA/cm^2 at AM1.5	(Lee, et al., 2012)
p-$GaInP_2$	-	Pt	2 M HBr, 2 M $NaClO_4$ (pH 0)	35 mA/cm^2 with W halogen	(Khaselev & Turner, 1999)

STABILITY CONSIDERATIONS

So far the photoelctrodes have been studied based on their potential to absorb light, generate and separate charge carriers, and inject them into the electrolyte to drive the reaction. This was then translated into actual efficiencies that could be measured. However, there is one more factor that would determine if a semiconductor is suitable for PEC applications: stability. This is one of the main factors to have into account when choosing a semiconductor material to be used as photoelectrode.

Although both the thermodynamic and kinetic properties play a role on the corrosion of a device, even slow corrosion can cause important damage to avoid practical use of a material. Thus, the thermodynamic properties of the reaction tend to be the ones considered more closely when looking at corrosion of a photoelectrode. Pourboix diagrams are an example of these properties, where a relation between the stable phase at a certain potential and pH is given. (Li, et al., 2013) Figure 9 depicts a case study of such Pourboix diagrams for TiO_2, The HER and OR are represented by blue dotted lines, and it should be noted that both are fully in the TiO_2 region, indicating that these reactions would happen at a potential where the semiconductor materials is stable as TiO_2. However, if the potential given is significantly lowered, the material might suffer from a reduction reaction and become unstable.

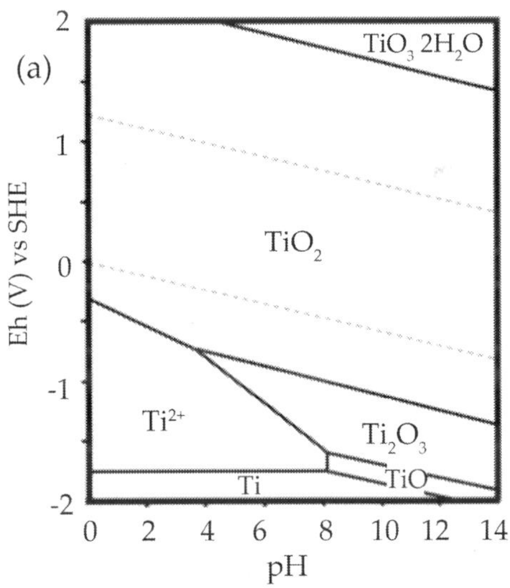

Figure 9: Pourboix diagram for titanium in aqueous electrolytes. Both H_2 evolution and water oxidation occur in the passivating region of the diagram (blue lines). From (Li, et al., 2013)

The morphology of the crystal surface also plays a role in the corrosion of the material, since corrosion is believed to happen only in certain crystal planes. In addition, the recombination at the surface is also important, since it can create defects and a starting point for the corrosion mechanism. Thus, in some cases, etching the surface of the photoelectrode can lead to a cleaner surface with less defects, increasing the performance and stability. However, the effect of this treatment would depend on the conditions at which it occurs and the surface morphology itself. (Memming, 1998)

In addition, the bandgap of the semiconductor materials must be considered, with semiconductors with bandgaps lower than 2 eV showing higher rates of anodic decomposition. The overall reaction of decomposition are both driven by thermodynamic and kinetic mechanisms, and a competition between the decomposition and the water splitting reactions is established. Only in very few cases could the decomposition reaction be completely suppressed to negligible values in comparison to the redox reaction. The transition metal chalcogenides such as WSe_2 are one exception, with high stability due to the position of the electronic states at the edge of the valence band, where non-bonding d-electron states of the metal are formed. Thus, metal scan sometimes act as barriers for anodic decomposition. However, this depends on the solution conditions, the metals used and its adhesion to the semiconductor. (Memming, 1998)

In summary, very few of the materials presented here combine a good performance regarding the water splitting reaction with long stabilities. In general, p-type semiconductors tend to be more resistant to corrosion. However, their performance is rather poor, with high overpotentials needed. This is generally attributed to high surface recombination and trapping. In that case, often precious metals need to be used as catalysts, which would increase the overall cost of the PEC device. (Memming, 1998)

Alternatively, an additional layer could be deposited at the interface between the semiconductor material and the electrolyte. This would change the band alignment to a more favorable position regarding the degradation redox reaction, and thus the stability of the overall device would be improved. (Guijarro, et al., 2015) These layers and the potential materials to be used are reviewed here.

Protective Layers

One of the most common and effective solutions to corrosion is the inclusion of a protective layer. Such layers could be metal (Ti) or other compounds (TiO_2). In addition, organic and polymeric layer could also be used as a protective layer attached to the surface by covalent bonds. However, it must be taken into account that the layer should not only protect the photoelectrode, but also work together with the semiconductor towards an efficient water splitting reaction. Thus, they should allow the charges to move through, and they should not block the solar light that drives the reaction. (Guijarro, et al., 2015) However, in many cases, the passivating layer is not electrically conductive, leaving the region protected but inactive. On the other hand, if the passivating layer acts as a sacrificial layer that dissolves in the solution, this would result in a more instable system, since one the layer is completely dissolve, the corrosion of the underlying layer will start. (Hu, et al., 2015)

Passivating layers that are electrically conductive, electrochemically stable and optically transparent are thus the preferred type of layer to deposit on relatively unstable photoelectrodes. In this respect, not only the material plays an important role, but also the deposition method, since it would determine the quality and crystal orientation of the final layer. Some of the most promising materials for both photoanodes and photocathodes are summarized in Table 7. Notice that most of them are metal oxides, such as TiO_2 or Al_2O_3. (Hu, et al., 2015)

Table 7: Summary of the most promising layers to be applied on both photocathodes and photoanodes. Adapted from (Liu, et al., 2014)

Working electrode	**Passivation layer**		**Electrolyte**	**Original stability**	**Stability after protection**	**Reference**
	Material	**Thickness (nm)**				
n-Si	TiO_2	2	1 M H_2SO_4	A few minutes	8 h	(Chen, et al., 2011)
n-Si	TiO_2	2	1 M Na_2HPO_4/ NaH_2PO_4 (pH 7)	< 3 h	18 h	(Chen, et al., 2011)
n-Si	TiO_2	2	1 M NaOH	< 0.5 h	8 h	(Chen, et al., 2011)
n-Si	Graphene	Monolayer	50 mM $K_3[Fe(CN)_6]$/ 350 mM $K_4[Fe(CN)_6]$	100 s	1000 s	(Nielander, et al., 2013)

n-ZnO	TiO_2	0.5-0.7	1.1 M KOH (pH 13)	-	> 3 h	(Liu, et al., 2013)
n-ZnO/p-Si	TiO_2	20	0.25 M $NaSO_4$ (pH 7.25)	A few minutes	24 h	(Kargar, et al., 2013) (Kargar, et al., 2013)
p-Si	TiO_2	100	1 M $HClO_4$	-	72 h	(Seger, et al., 2013)
(p-i-n)a-Si	TiO_2	80	0.5 M KHP (pH 4)	12 h (10 %)	12 h(95 %)	(Lin, et al., 2013)
p-Cu_2O	TiO_2	11	1 M Na_2SO_4 – 0.1 M K_2HPO_4 (pH 4.9)	< 5 min	1 h	(Paracchino, et al., 2011)
p-Cu_2O	Carbon layer	20	1 M Na_2SO_4	20 min (12.6 %)	20 min (87 %)	(Zhang, et al., 2013)
p-Cu_2Zn-SnS_4	TiO_2	A few	0.5 M KH_2PO_4-K_2HPO_4 (pH 7); 0.1 M Na_2SO_4 NaOH (pH 9)	Within seconds	15 min	(Rovelli, et al., 2013)
p-InP	TiO_2	2-5	1 M $HClO_4$ (pH 0.51)	4 h (67%)	4 h (~100%)	(Lee, et al., 2012)

The coverage of this layer is also relevant, since the semiconductor should not be in direct contact with the electrolyte. The protective layer must be as conformal, homogeneous and free of pinholes as possible, not to expose the semiconductor material underneath to the corrosive medium. This has to be taken into account also when producing nanopatterned photoelectrodes. Thus, the technique used to deposit these layers plays a crucial role. Physical-vapor deposition (PVD) can achieve good properties, but the technique also leads to inhomogeneities that might expose some of the semiconductor materials to the electrolyte Chemical vapor deposition (CVD) techniques, and in particular atomic layer deposition (ALD) show a better behavior, since they are able to deposit very homogeneous and conformal layers. (Hu, et al., 2015) However, it must be noticed that these are in general time consuming techniques, leading to higher costs in the PEC device, as would be commented more in detail in future sections. Thus, there is a trade-off between a highly stable material and the cost of such material.

Finally, it is important to notice that a protective layer can not only improve the corrosion behavior of the photoelectrode, but also the band bending at the surface, thus improving the PEC performance of the device. In addition, it is possible that these films have catalytic properties, which further enhance the performance of the devices. (Hu, et al., 2015)

CATALYST MATERIALS

The last element to be considered in the PEC device is the catalyst material. As previously discussed, catalyst materials are deposited at the semiconductor/electrolyte interface in order to improve the kinetics of the reaction, facilitating the water splitting reaction and reducing the required overpotentials. Thus, it is important to select the correct catalyst for each PEC device. The catalyst activity can be divided depending on the targeted half reaction: the hydrogen evolution reaction (HER) and the oxygen evolution reaction (OER).

HER

As previously mentioned, the two half reactions that compose the water splitting reaction would need different electrodes, and the characteristics that would make good catalyst can be determined by plotting the Tafel plot. The catalyst used in the HER are summarized in Figure 10. The most efficient and active materials would be Pt or Pd, but they are rare and expensive metals, and thus it is interesting to further look for alternatives. Aside from the pure metals considered previously, a variety of other, more cost-effective materials can be considered. These are mostly either alloys incorporating non-metallic elements, such as NiMo, Ni-Co or more complicated mixes such as Ni-Mo-Fe; or metal oxides such as RuO_2, $SrNbO_3$, sulfides. However, since platinum and other noble metals have also been used as demonstrators fort he HER, allowing for very high efficiencies, a trade-off between price and efficiency would be established. (Walter, et al., 2010)

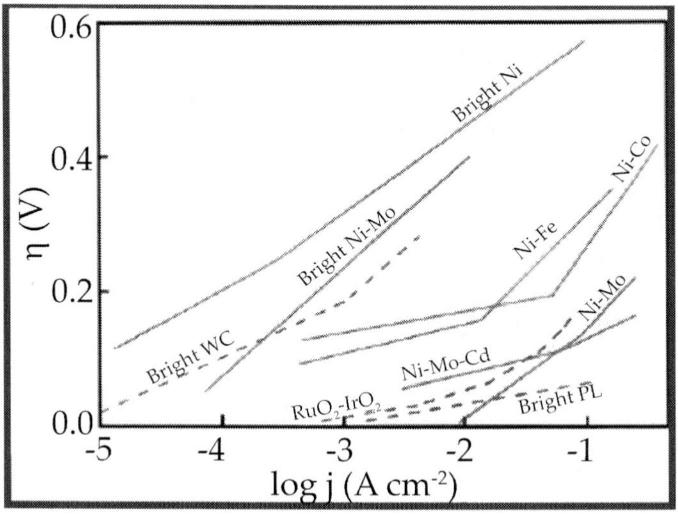

Figure 10: Tafel plots for the following HER catalysts, reproduced from the associated references: bright Ni, Ni-Mo-Cd; bright Ni-Mo; Ni-Fe, Ni-Co, Ni-Mo; bright WC; RuO_2-IrO_2; bright Pt. "Bright" refers to electrode surfaces that have been polished such that the electrochemically active area equals the projected area, whereas electrodes not noted as "bright" have not been corrected for electrochemically active surface area. Solid red curves were collected in basic electrolyte, whereas dashed blue curves were collected in acidic electrolyte. From (Walter, et al., 2010)

There are several factors to consider when depositing the catalyst, beside the material itself. First, it must be noted that the catalyst would be deposited on top of a photoelectrode, and thus it would need to be able to adhere to this material and it should not affect the semiconductor itself by hard deposition conditions. (Khnayzer, et al., 2012) Moreover, the crystallinity and crystal orientation can also affect the resulting performance of the photocathode. (Yang, et al., 2017) Thus, not only the material, but also the crystallinity and structure of the material needs to be considered.

OER

In this case, the catalyst plays an even more crucial role, due to the complicated kinetics of this half reaction, since it needs four electrons for each oxygen molecule produced, as opposed to the two electrons needed for HER. This results in relatively high overpotentials, and difficulties in describing the performances of the catalysts used. The most used an well-known catalyst for this reaction is RuO_2 under acidic conditions, and $NiCo_2O_4$ for alkaline conditions. In these cases, as Figure 11 shows, the required overpotentials are drastically increased, with 300 mV for the $NiCo_2O_4$ as an example.

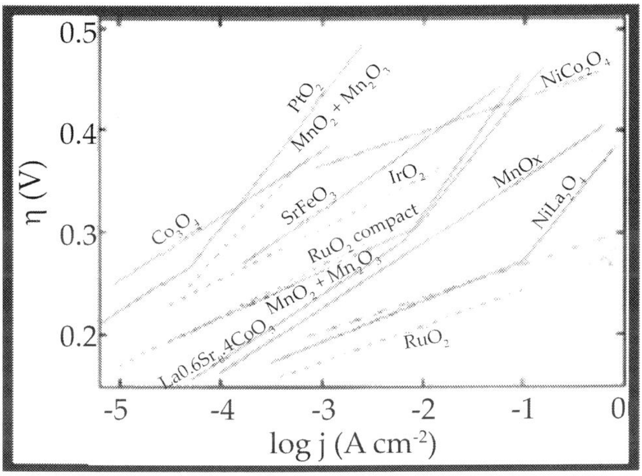

Figure 11: Collected Tafel plots for OER catalysts, PtO_2; MnOx; Co_3O_4; $NiCo_2O_4$; $SrFeO_3$;IrO_2; RuO_2 (both "compact" and not); La0.6Sr_0.4CoO_3;NiOX; $NiLa_2O_4$. Solid line$_s$ represent alkaline conditions, and dashed lines are under acidic conditions. Current densities are in terms of independently measured electrochemically active surface area. The dot-dashed line, however, shows $NiCo_2O_4$ with respect to projected, rather than active surface area, demonstrating that the apparent activity of a catalyst with a high roughness factor increases dramatically if one does not account for the catalytic activity with respect to the active surface area. From (Walter, et al., 2010)

Another important factor to take into account when considering the catalyst performance, as in the previous case, is the preparation techniques, which could greatly affect the resulting overpotentials. That is explained both by the nanostructures formed and by the orientation and crystallinity of the metal, which can also greatly affect the activity of the catalyst. Regarding the materials that could be practically used in this half reactions, as depicted in Figure 12, RuO_2 and IrO_2 give an efficient but expensive alternative, while inexpensive metal oxides TiO_2, SnO_2, Ta_2O_5 could also be used with reasonable results. However, if alkaline conditions can be applied, nickel and its alloys would be the materials of choice. (Walter, et al., 2010)

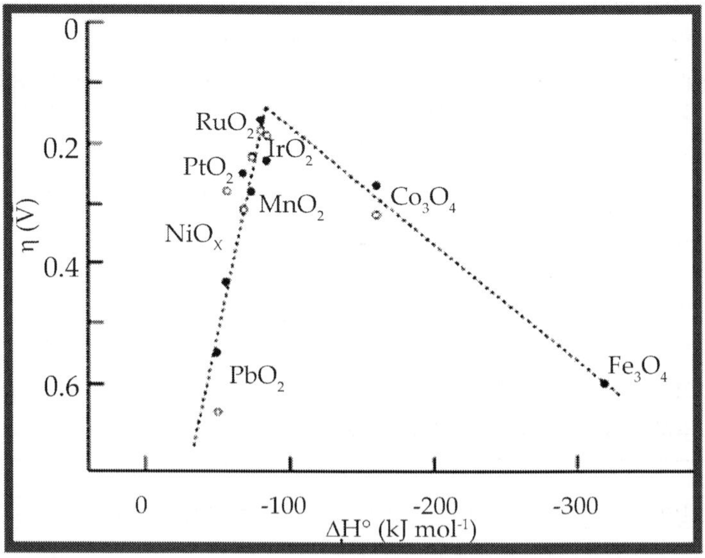

Figure 12. OER volcano plot. Closed circles: in acid. Open circles: in alkaline media. From (Walter, et al., 2010)

SUMMARY

There is a wide range of materials that can be used in a PEC device based on the necessary conditions outlined in the previous section. This section has explored the most promising materials both for photoanodes and photocathodes, as well as possible and protective layers and catalysts of both reactions.

Regarding the materials for both photoanodes and photocathodes, there seems to be two main options: metal oxides and III-V materials. The difference between the ones used for HER and OER is the band alignment, and thus OER would happen on the surface of n-type semiconductor materials such as TiO_2, WO_3 or GaAs, and HER would be on p-type semiconductor materials such as CuO_2 or InP. The performance of the PEC device would be higher in III-V materials, since the charge carrier generation, separation and collection is increased by the high quality material. However, since the fabrication process are rather time consuming, and the elements used are rare, these materials are not practical to be applied in a practical PEC device. Alternatively, metal oxides and other silicon based materials, with lower performances and often lower stabilities, show a possible option. However, these materials would need other supporting materials to be effective and durable, such as a protective layer between the semiconductor and the electrolyte, and a catalyst to reduce the energy barrier for the reaction to occur.

First, regarding passivating layers, they have been reviewed extensively lately, finding that according to both Pourboix diagrams and stability experiments, TiO_2 seems to be the most promising material to use as a protective layer. In addition, it was shown that, since this is an n-type material, it will not only improve the stability of the device, but it would also create an electric field that would use the favorable band alignment to more effectively separate and inject charge carriers, yielding a higher efficiency.

Finally, the possible catalysts were considered. The same trade-off appears when considering these materials, with the best and most stable performances being achieved by precious metals such as Pt and Pd. However, other materials can also achieve respectable materials, mostly formed by metal alloys such as Ni-Mo, Ni-Co or Ni-Mo-Fe. It is important in any case to account not only for the materials used, but also the electrolyte and pH, as well as the deposition method. Thus, these parameters should be further explored in following chapters and deeper investigate the effects that each of these would have in the performance of the PEC devices.

CHAPTER 4

EMERGING TRENDS

PEC devices suffer from a wide variety of losses, including overpotential losses at the anode and cathode interfaces (activation energies of the reactions, kinetics, mass-transport of the multistage half-reactions, etc.), the ionic conductivity overpotential in the electrolyte, and several solid state balance at the semiconductor. These losses are substantial, adding from a few hundreds of millivolts to almost one volt as overpotential, depending on the catalysts, electrolytes and semiconductor materials used. Thus, the final voltage needed for the reaction to happen in a stand-alone setup would be between 1.6 to 1.9 V. (van de Krol & Gratzel, 2012) Thus, the PEC process needs a delicate balance between the different factors.

To improve the STH efficiency of these devices, the losses must be minimized by applying different strategies. These must address all the main losses of the systems, including optoelectronic losses, catalyst losses, and ion transport in the electrolyte. Thus, the bandgap of the material has to be finely chosen and tuned for maximum light absorption, as well as considering the thickness and texturing and the direct/indirect bandgap. In addition, the interfacial charge carrier extraction and product formation can be improved by changing the interface conditions, such as band alignment, surface area or surface kinetics. Including nanoparticles and nanopatterns in this case might aid the surface kinetics. However, these approaches also increase the surface recombination losses, and thus further optimization is needed to reduce the losses. Moreover, the charge carrier extraction and transport in the electrolyte can be improved by changing its composition. Some studies in fact use seawater as the electrolyte instead of distilled water. Other studies would use membranes to be able to separate the two sides of the electrodes, and thus regulate the conditions separately. Finally, the distance between electrodes can be optimized. (van de Krol & Gratzel, 2012) In this chapter, some of these challenges and innovative solutions are explored in more detail.

NANOSTRUCTURES AND NANOMATERIALS

In recent years, the use of metal oxides has been looked at more closely, as both efficiencies and stabilities of these materials have improved. However, the charge carrier mobility and carrier lifetime is still low due to the great defect density in these materials and the possibility of obtaining non-stoichiometric materials, which leads to dangling bonds. Thus, the distance that the charge carriers must travel has to be minimized. To achieve this, nanostructures can be used to minimize recombination and increase surface area. Recently, new nanostructuring patterns and approaches have emerged, suggesting that it would be possible to solve these problems. (Warren, 2011) In addition, these nanostructures increase the exposed area to the solution, increasing the active total area, and thus the efficiency of the devices. Finally, these structures would cause scattering of light, and thus more efficient light absorption within the semiconductor, which would also contribute to enhancing the STH efficiency of the device. (Li & Zhang, 2010)

To understand the different approaches here explained, first a brief account of the principles behind this approach need to be understood. A central idea within charge carrier mobility improvement by applying nanostructures is that of a polaron. When charge carriers are added to an uncharged lattice, this charge tends to delocalize over a certain number of atoms. The polaron represents the area over which these charge carries spread, and which can be over many atoms (large polaron) or localized over a few atoms (small polaron). These charges would affect the bond lengths and angle, especially when they are much localized within the lattice, inducing changes in the electron density. The polaron links together the appearance of charge in a certain area and the lattice distortion, reducing it to a quasi-particle. This is also connected with the concept of self-trapping. When a charge carrier is localized in a small polaron, the lattice shifts to a new equilibrium within a period of a few lattice vibrations. This stability increase creates a deeper potential well that would need to be overcome to transfer the polaron into a neighboring atom. Thus, the activation energy is affected, where small polarons are associated with lower drift mobility than large polarons. Thus, there is an estimated upper theoretical limit for polaron mobility of 10 cm^2/V s for large polarons and 0.1-0.01 cm^2/V s. Another characteristic is that the mobility of small polarons increases with temperature because then the transport is aided by lattice vibrations, while in large polarons it decreases with temperature since the lattice vibrations scatter the carriers. A lower mobility would imply that the time required for a minority carrier to cross the space charge region is increased, with more chances for recombination. That means that small polarons need higher band bending and narrower space charge layer widths for charge separation, and thus higher doping levels. In this structure, it is less likely for light to be absorbed in the space charge layer, reducing parasitic absorption. However, this changes if non-planar geometries are explored, specifically if the semiconductor fits within the space charge layer, it should be possible to achieve more efficient devices, since in that case the high doping does not significantly affect the minority charge carrier lifetime. However, the mid-bandgap states still have an effect in the recombination rate. (Warren, 2011)

In conclusion, semiconductors with small polaron transport and short lifetimes can be used to make efficient device when some conditions are met. First, doping levels should be high, to eliminate recombination within the depletion region. Second, light absorption should be confined to the space charge layer, which is unlikely for electrodes with planar geometries, but it is possible with nanostructures.

Some structures have been designed to meet these criteria, (Warren, 2011) as shown in Figure 1. These mainly consist of ordered structures and nanotubes of different dimensions in terms of length, diameter and wall thickness. To further explain how these properties affect the PEC device and semiconductor materials within it, each of these cases is studied individually.

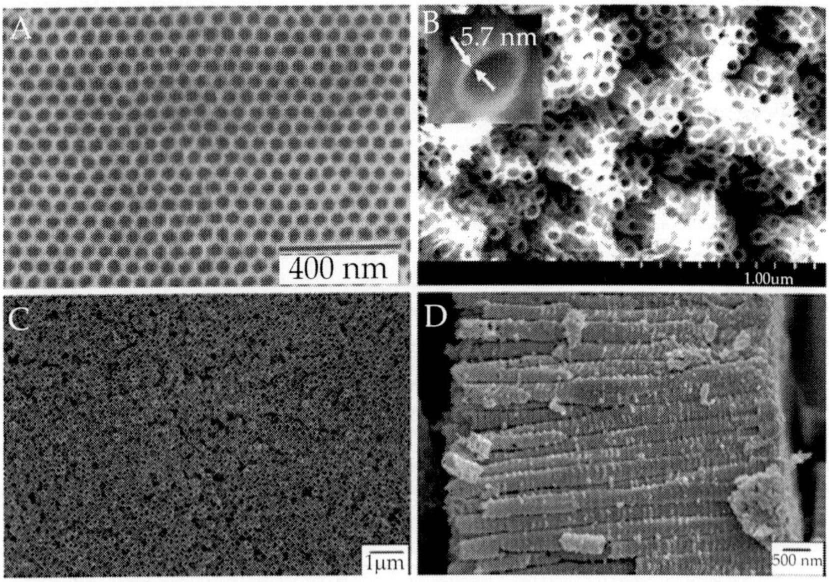

Figure 1. SEM images of the anodization products. a) Large scale ordered aluminum oxide made by anodization in sulfuric acid, b) hematite nanotube arrays prepared by sonoelectrochemical anodization in ethylene glycol, and a single nanotube shown in the inlet, c) titania nanotube arrays prepared by anodization in aqueous electrolyte, and d) side view of the same tubes. From (Warren, 2011)

Nanopores and Nanotubes

Anodization of certain metals leads to porous structures of metal oxides and hydroxides, including Al, Ti, Ta, Cd, Nb, Mg alloys, W, Sn, Fe, Ag and Si. For example, anodization of Ti leads to the production of TiO_2 nanotube arrays. Moreover, anodization of iron would lead to nanoporous and nanotubular oxide structures. Pores in the range of 20-100 nm are typical for these structures and materials; for the nanotubes, the wall thickness is in the range of 5-30 nm. (Warren, 2011)

In the context of water splitting, TiO_2 nanotubes seem to attract the most attention due to the easy fabrication process and modest charge carrier lifetime. (Cho, et al., 2015) It is important here which crystalline structure the TiO_2 adopts (anatase and rutile), which would mainly depend on the fabrication technique and post-treatments applied. Moreover, studies in these materials suggest that the whole transport is controlled by small polarons. However, it must be taken into account that in these structures, grain boundaries, defects and other defects would still increase recombination and reduce the charge carrier lifetime. (Grimes & Mor, 2009) reported the application of these nanotubes for water splitting using long TiO_2 nanotubes with good IPCE values of over 90 %. Later on, it was suggested that the key aspect of the improved performance was that the space charge layer should fit entirely within the nanotube wall, as discussed previously, emphasizing the importance of the electric field based charge carrier separation. The best performing devices have achieved efficiencies of 16.25 %, although applying UV light instead of solar light. Due to the high bandgap of TiO_2, solar light will not excite great amounts of electron hole pairs, and thus the photocurrents produced would not be high, limiting the STH efficiencies. Doping can be used to enhance light absorption, but it also creates localized states within the bandgap, reducing the overall efficiency. (Warren, 2011) Thus, other materials must be studied to prove the applicability of nanotubes and nanoarrays on PEC devices driven by solar energy.

Iron oxide is another well-known material benefiting from nanostructuring approaches. In particular, hematite is a small polaron conductor, although the hole transport is anisotropic and thus the polaron might be bigger in a certain direction. (Deng, et al., 2014) Dopants would also play a role on this anisotropy, allowing for tuning of these properties. However, charge carrier lifetimes in hematite are much shorter than in TiO_2, needing much smaller features in their nanostructures to achieve similar properties, with dimensions of less than 10 nm. (Warren, 2011) These approaches have successfully been implemented, with current densities of 0.4 mA/cm^2 at 1.5 V vs. RHE. (Wang, et al., 2015) Moreover, heat treatment can help with the crystallization of the hematite structures, reaching higher efficiencies. (Sivula, et al., 2010) Finally, the efficiency can be improved by creating more vacancies in the structure, which act as a sort of doping. This is achieved by performing the nanotube formation under a reducing environment. (Lin, et al., 2011)

Nanophotonics

In addition to improving the structure and minority charge carrier mobility, nanostructures can also improve the light management and absorption. This area searches to manipulate photons using nanostructures for more efficient light management, opening the possibility of reducing the amount of material needed for efficient light absorption. There are several emerging areas within this field, including plasmonics, multiple exciton generation, photonic crystals and up-conversion. However, not all of these strategies can be applied directly to water splitting. Among the most useful strategies to apply in this field are plasmonics. This idea consists on the fact that certain nanostructured materials can confine and manipulate or scatter single photons. In particular, photoelectrochemical water splitting would be interested in near-field absorption and light scattering, in order to increase the light absorption in the semiconductor material. (Warren, 2011)

When a metal nanoparticle is subjected to an oscillating electric field, such as light, free electrons in the nanoparticle will respond to the field by oscillating, and part of the light would be polarized. The light used to create these resonances would depend on the metals used, and although most metals have these resonances in the ultraviolet range, in particular gold and silver are more polarizable in the visible or infrared part of the light spectrum. This will cause a non-random displacement of electrons, which produces an electric field that extends further than the nanoparticle, and into the semiconductor. Thus, this is able to create an electron-hole pair if the energy of the plasmon is higher than that of the semiconductor bandgap. Since the electric field effect reduces further away from the metal particle, this effect is called near-field absorption. Moreover, it is known that the behavior of the resulting plasmon can be tuned by changing the metal particle size. These effects have been modelled, as shown in Figure 2, where the absorbance enhancement effects can be clearly seen surrounding the metal nanoparticles, in this case based on silver. (Warren, 2011)

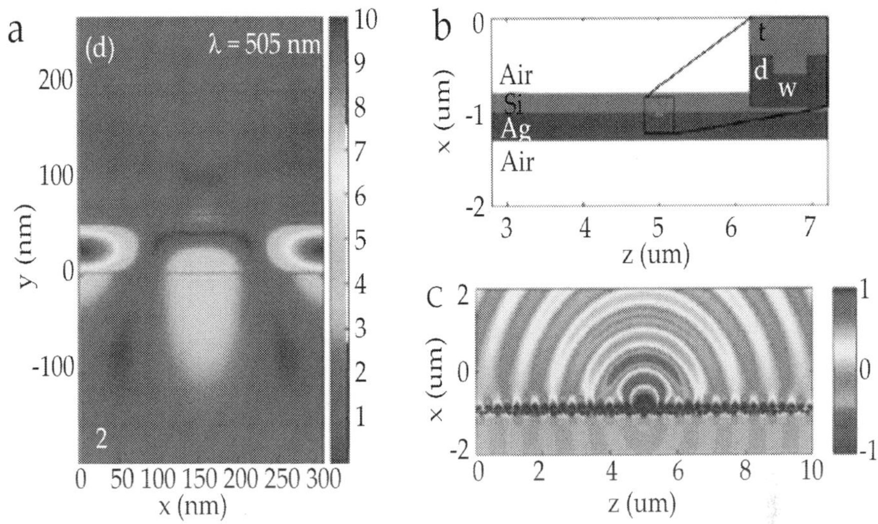

Figure 2: Models of scattering from metal nanostructures. a) Showed at a certain particle geometry, b) model of a Si/Ag device with 100nm wide and 50 nm deep grove and c) simulation of the field scattered in such a grove. From (Warren, 2011)

Despite the optical potential of this approach, there are other factors to take into account when introducing this concept into a PEC device. When these particles are introduced at the semiconductor/electrolyte interface, the stability, Fermi energy levels and band bending produced must also be taken into account. Other effects that could be observed are the catalytic activity of the metal nanoparticles and the possibility of creating trap states at the surface, which may lead to recombination. Thus, even though introducing nanostructures and nanoparticles in PEC devices could have positive results, they might also lead to other problems that might reduce the overall STH efficiency. (Warren, 2011)

Quantum dots (QDs)

These are small particles or crystals embedded in a material matrix. These structures are able to effectively tune the bandgap of the semiconductor as a whole by varying the QD size. Smaller QDs would yield higher bandgaps, while bigger ones will have smaller bandgaps. Thus, by using QDs it is possible to sensitize a certain material, such as TiO_2, for light absorption, substantially increasing its response to visible light. (Li &

Zhang, 2010) However, it must be mentioned that this approach might also lead to multiple grain boundaries, which are powerful recombination centers for charge carriers. This would increase the recombination rate, lowering the obtained STH efficiencies from these kind of devices.

The crystal diameters usually range between 1-20 nm, and due to their small size, they can benefit from the effect of quantum confinement. This is mainly an optical effect, and thus these materials are often considered for optical devices such as LEDs, solar cells, etc. (Trinidade, 2001) An advantage of these materials is that, by tuning the size of the crystals, the bandgap can be chosen. In general, smaller crystals will have higher bandgaps, while bigger crystals will have lower bandgaps. (Anikeeva, et al., 2009) This is illustrated in Figure 3.

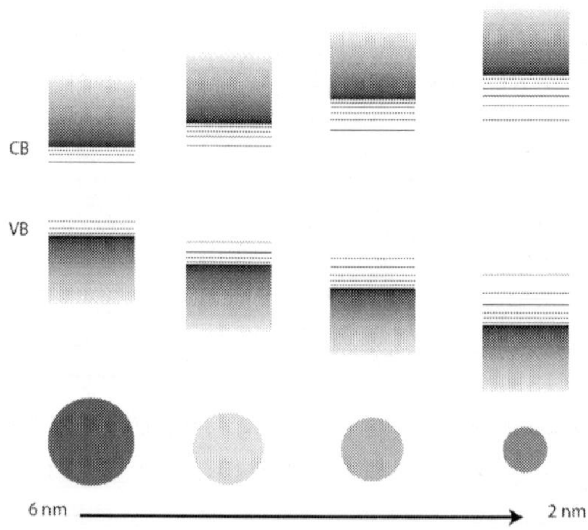

Figure 3: Schematic of the band diagrams of the material related to the quantum dots size in the matrix, and the associated colors to each band gap

Nevertheless, it must be taken into account that when nanocrystals are included in a material matrix, the electrical properties would be affected. A great amount of grain boundaries are introduced in the material, and thus the recombination rates would be increased significantly. Thus, some passivation steps must be introduced in order to mitigate the effects of grain boundaries. (Alivisatos, 1996) In addition, the deposition methods for these materials at the moment are usually rather time consuming, reducing the industrial throughput and cost effectiveness. That is an issue that should also be addressed for large-scale implementation. (Kovalenko, 2015) An approach that has been proposed to overcome some of these issues is to use these quantum dots as catalysts, (Yeh, et al., 2014) although some of these issues still remain to be further improved.

MULTIJUNCTION APPROACHES

As previously explained, PEC devices consists of a semiconductor, which would absorb light to drive a half reaction. Thar could either be a photoanode or a photocathode, as shown in Figure 4a and b. However, because of the limitations in bandgap and light absorption that a single material poses, multijuncion approaches are often used, as depicted in Figure 4c. In that case, several semiconductors would be connected in series to achieve a high enough voltage to drive the redox reaction, and more efficiently utilize the solar spectrum. (Ager, et al., 2015)

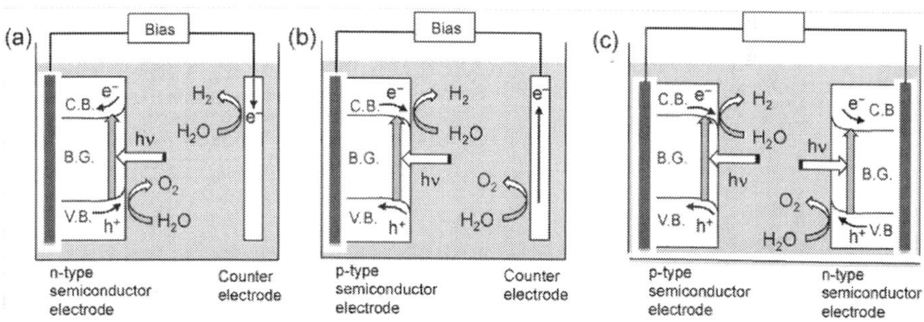

Figure 4. Schematic diagram of the photoelectrochemical cell assemblies used for hydrogen production. (Ager, et al., 2015)

In fact, there are several ways to create multijunction devices for water splitting. Among the most promising ones are the combination of a photoanode and a photocathode in one device, the combination of a PEC device (both photocathode and photoanode) and a photovoltaic (PV) cell, or the use of a full multijunction PV cell with the addition of a catalyst. In this section, each of these options will be investigated in detail.

However, before diving into the different typologies it is important to describe what is meant by a PEC device and a PV device. When looking at multijunction devices especially, the line between PV and PEC seems blurred. (Nielander, et al., 2015) proposed a flowchart, shown in Figure 2, with a detailed recommendation on how to identify and classify the different possible devices combining PV and PEC. For the scope of these book, a more simple classification will be used, dividing the devices into purely PEC, PEC with a PV cell attached, and purely PV combined with some sort of electrolysis system. Purely PEC devices depend only on the semiconductor/electrolyte interfaces to enhance the charge collection and injection, whereas the PV devices have an in-built junction in the semiconductor material that allows for charge carrier separation and collection independently from the electrolyte.

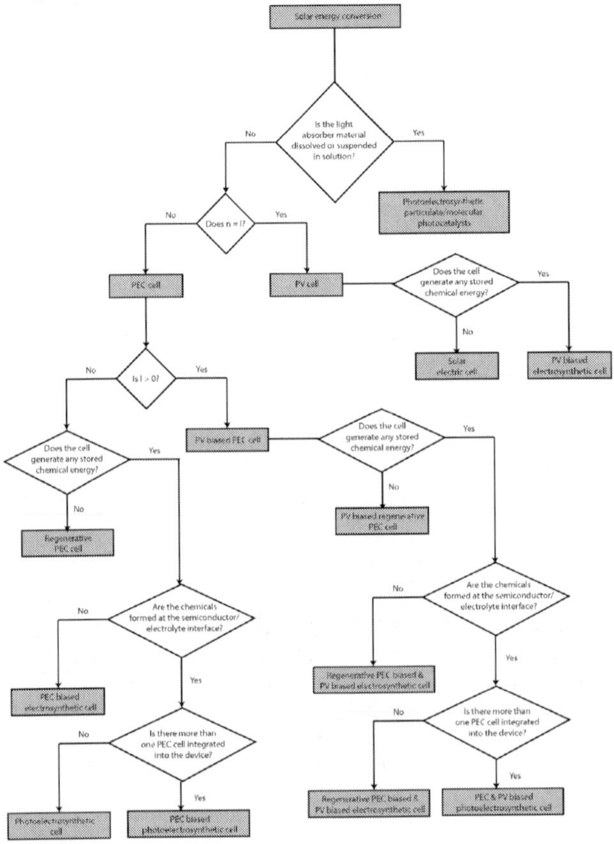

Figure 5. Proposed classification for solar energy converters based on the different semiconductor and electrolyte junctions, as proposed by (Nielander, et al., 2015)

PEC devices in series

This concept was already experimentally proved in 1975, with a device consisting of a p-GaP/n-TiO_2 tandem combination and no external bias able to drive the HER and OER. (Yoneyama, et al., 1975). However, the efficiencies of this device are still low. Since then, several groups explored this concept with a combination of III-V technologies and metal oxides that were able to produce higher STH efficiencies than a single photoelectrode. The two main challenges of this devices in the early days were found to be the relatively low STH efficiencies and the stability of the photoanode. (Ager, et al., 2015)

To look at the main devices and points of improvements for these devices, the fundamental principles have to first be understood. The functioning principles of such a device are shown in Figure, where the band diagram shows what it is often called a buried junction. In this case, the interface between the two materials plays a crucial role. Since the charge carrier flow should be the same in both semiconductor materials, the charge carriers would need to recombine in the tunneling recombination layer (TRJ) between the two semiconductors, to create a continuous flow of current.

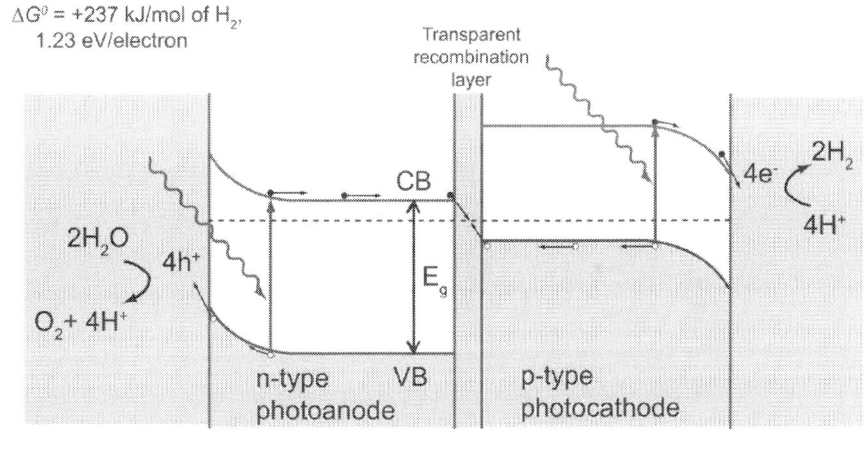

Figure 6. Schematic of an idealized tandem-junction photoanode and photocathode device during steady-state operation. The process of solar water splitting is overlaid on the equilibrium diagram. Proton conduction in the electrolyte from the anode to the cathode is required for continuous operation. From (Ager, et al., 2015)

When looking at this junction, there are then two main components to focus on: the semiconductor/electrolyte junctions and the buried junctions between the two different electrolytes. In Figure 7 the different possibilities are presented, where it can be seen in a) that the Fermi level splitting happens at the interface between semiconductor and the electrolyte, while in b) this splitting happens at the interface of the two semiconductors, in the so-called buried junctions. That would depend on the couple of semiconductors chosen, as well as the tunnel recombination junction properties. Moreover, this figure also show the case in which either the photoanode or the photocathode would receive the light first, which would depend mainly on the bandgap of each of the semiconductors, placing the higher bandgap material in front, and the lower bandgap semiconductor behind, since it would be able to absorb the light that it would not be absorbed in the first.

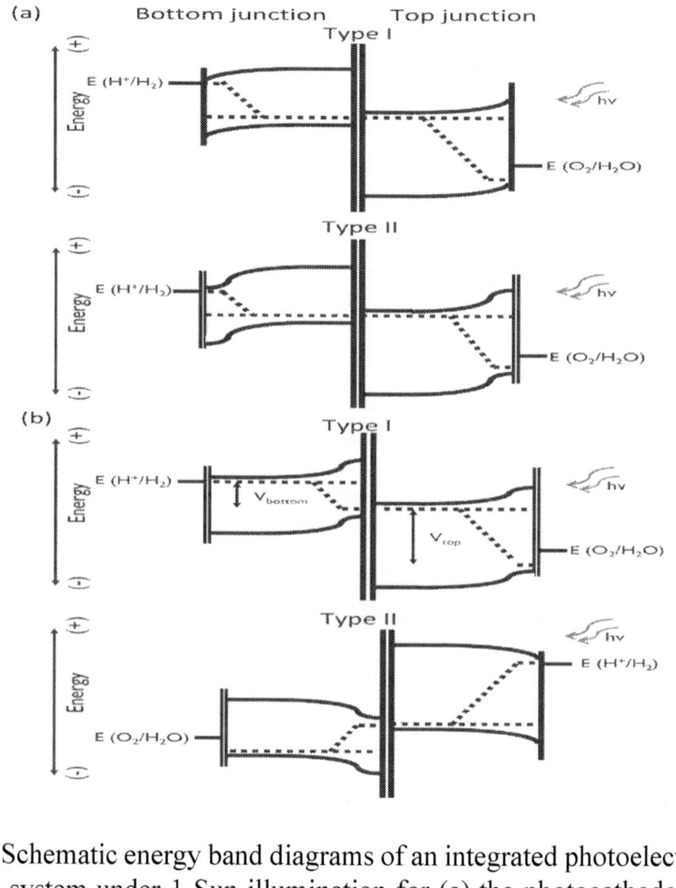

Figure 7. Schematic energy band diagrams of an integrated photoelectrochemical (PEC) system under 1 Sun illumination for (a) the photocathode + photoanode PEC system, denoted as Configuration A, and (b) the tandem light absorber + electrocatalyst PEC system, denoted as Configuration B. Types I and II illustrate the semiconductor–liquid junctions and buried junctions, respectively, with an Ohmic contact. The photogenerated minority-carrier electrons drift and diffuse to the photocathode–electrolyte interface and reduce H^+ to H_2, whereas the photogenerated minority-carrier holes drift and diffuse to the photoanode–

electrolyte interface and oxidize water to O_2. The majority carriers (holes in the photocathode and electrons in the photoanode) recombine at the Ohmic contact, which is indicated by the doubly thick line that connects both photoelectrodes back-to-back. From (Hu, et al., 2013)

All these parameters would affect the performance of the combined device. However, it is difficult to directly use these characteristics to predict exactly how much this performance would be affected. In order to know the suitability of the chosen combination, there is a more direct method to predict the resulting current density by comparing the two voltage-current density characteristics of each photocathodes, as shown in Figure 8. Obviously, this would not be the final performance, since the bottom electrode would absorb less light when in a tandem device. Nevertheless, it would still give an idea of how it would perform.

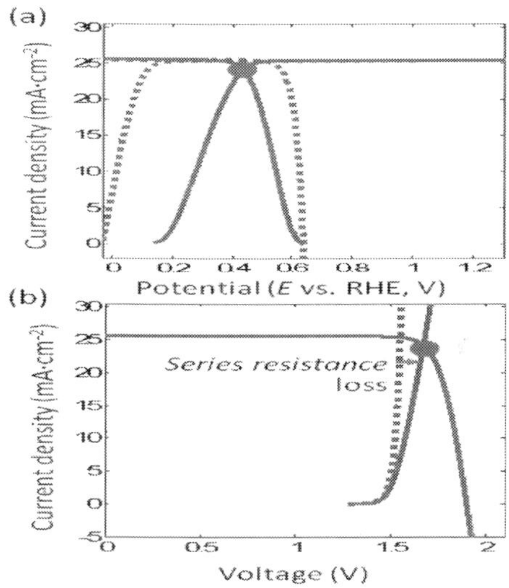

Figure 8. Operation points (red dots) of (a) a photocathode + photoanode PEC configuration, and (b) a tandem absorber + electrocatalyst PEC configuration. The dashed blue and green curves in (a) are the ideal J–V properties of the various photoelectrode materials. The dashed curve in (b) represents the electrochemical load without solution resistance losses. For (a) and (b), Pt and RuO2 were chosen as the HER and OER catalysts, respectively, the light absorber had FF 0.85, and the solution resistance was 5 ohm cm^2. From (Hu, et al., 2013)

Looking at the possible materials, most of them rely on III-V technologies, especially in the earlier stages. A dual bandgap p/n PEC was later reported using Fe_2O_3 and p-GaP, with RuO_2 and Pt as catalyst respectively, with STH efficiencies around 0.1 %. An important factor from this work is the identification of the Fe_2O_3 quantum efficiency as the limiting factor. (Mettee, et al., 1981) By using InP, the STH efficiency of the complete device can be dramatically increased to 8.2 %. The most remarkable devices made to date are summarized in Table 1.

Table 1. Two junction tandem PEC cells water splitting demonstrations with at least one semiconductor-liquid junction. Adapted from (Ager, et al., 2015)

Device structure (cathode/anode)	Conditions	STH efficiency	Stability	Reference
Cu_2O (p) / $BiVO_4$ (n)	KHP (pH 6), 100 mW/cm^2	0.5 %	2 min, 20 % current loss	(Bornoz, et al., 2014)
Cu_2O (p) / WO_3 (n)	0.1 M Na_2SO_4 (pH 6) , 100 mW/cm^2	0.04 %	Not reported	(Lin, et al., 2012)
InP (p) / GaAs (n)	6 M KOH	8.2 %	10 h, 10 % current loss	(Kainthla, et al., 1987)
GaP (p) / Fe_2O_3 (n)	1 M Na_2SO_4	0.02-0.1 %	Not reported	(Mettee & Calvin, 1981)
CdTe (p) / TiO_2 (n)	1 M NaOH, 100 mW/cm^2	0.044 %	1 h stable photovoltage, some cells tested to 50 h	(Ohashi, et al., 1977)
GaP (p) / TiO_2 (n)	1 M NaOH, 100 mW/cm^2	0.098 %	h stable photovoltage, some cells tested to 50 h	(Ohashi, et al., 1977)
CdTe (p) / $SrTiO_3$	1 M NaOH, 100 mW/cm^2	0.18 %	h stable photovoltage, some cells tested to 50 h	(Ohashi, et al., 1977)
GaP (p) / $SrTiO_3$	1 M NaOH, 100 mW/cm^2	0.67 %	h stable photovoltage, some cells tested to 50 h	(Ohashi, et al., 1977)
GaP (p) / TiO_2 (n)	0.2 N H_2SO_4, 85 mW/cm^2	0.25 %	Not reported	(Nozik, 1976)

Another crucial factor that was identified in this analysis was the bandgap combination between the two photoelectrodes. (Walter, et al., 2010) By modelling the Shockley-Queisser limit, the STH efficiency of several material combinations could be estimated. (Hu, et al., 2013) calculated the possible theoretical efficiencies and plotted them as a function of the bandgap of both materials. In such estimation, the photoelectrocatalysis has been assumed to be independent from the light absorption. These calculations also assume that defects are not present at the semiconductor/liquid interface, and that the barrier height of the semiconductor/liquid interface is close to the bandgap, having the right band alignment.

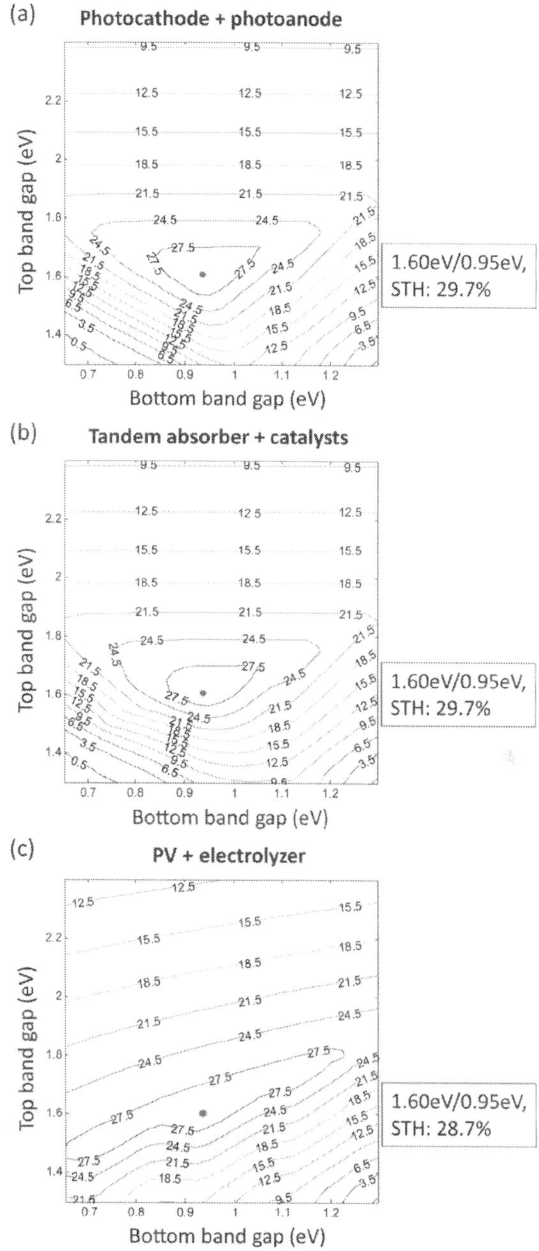

Figure 9. Iso-efficiency plots showing the STH efficiency limits for (a) a photocathode + photoanode PEC, (b) a tandem absorber + electrocatalyst PEC, and

(c) a two-junction PV + electrolyzer. In (a) and (b), Pt and RuO_2 were chosen as the HER and OER catalysts, the light absorber had FF ¼ 0.85, and the solution resistance was 5 ohm cm^2. In (c), the electrolyzer efficiency was taken to be 73%. From (Hu, et al., 2013)

However, potential ohmic losses were taken into account, as well as the Tafel slopes from the different common catalysts (Pt, Ni-Mo, RuO_2, $NiFeO_x$), to have a more accurate idea of the actual potential performance. The results of these simulations can be seen in Figure 9. From this study, the optimum bandgap combination has been determined for different characteristics. In addition, the potential performance of a PV direct system and PV indirect system have been plotted. It is remarkable that in the PEC and direct PV cases, the results are the same, since the possible differences such as band alignment were not considered in the simulations. However, the indirect PV system (PV + electrolyzer) show some differences with respect to the other two systems due to the losses assumed in the cables and external circuitry needed. Ultimately, according to the simulations by (Hu, et al., 2013), the maximum efficiency that could potentially be achieved is 29.7 % with a tandem device using a semiconductor with a bandgap of 1.6 eV combined with a material of bandgap 0.95 eV.

(Hu, et al., 2013) also considered the different resistances and fill factors that this device could have, as well as the use of other more earth-abundant and cost effective materials and studied how the optimum would vary with the different tendencies. This study is thus interesting to determine which characteristics should be further improved for higher STH efficiencies.

In the case of more earth-abundant catalysts, such as Ni-Mo and $NiFeO_x$, the efficiencies are lowered to 28.5 %. Interestingly, when changing the catalyst material, they determined that also the optimum bandgap would slightly change to 1.65 eV and 1 eV, respectively.

Another take on the PEC tandem devices is the so-called Z-scheme. This consist in applying the tandem concept to photocatalyst, having one OER catalyst and one HER catalyst in different n- and p-type particles, as shown in Figure 10. This approach gives some hope for a low cost solution to water splitting, especially considering the fact that, this way, the costs associated with noble metal catalysts would be reduced. However, it also has serious challenges to overcome. The charge carrier separation within each catalyst particles is a challenge, and new strategies need to be implemented to increase it. In addition, the particles should simultaneously dive two reactions in localized areas, which can be a challenge. Functionalized sites by using local catalyst could aid with it, but it would increase the price of the overall system. Finally, looking at the reactor scale, it this case both the hydrogen and oxygen are produced in the same particle bed, difficulting the task of separating the produced hydrogen and oxygen gases. A double bed system could be used instead, increasing the costs and mass use and diminishing the ion transfer. (van de Krol & Gratzel, 2012)

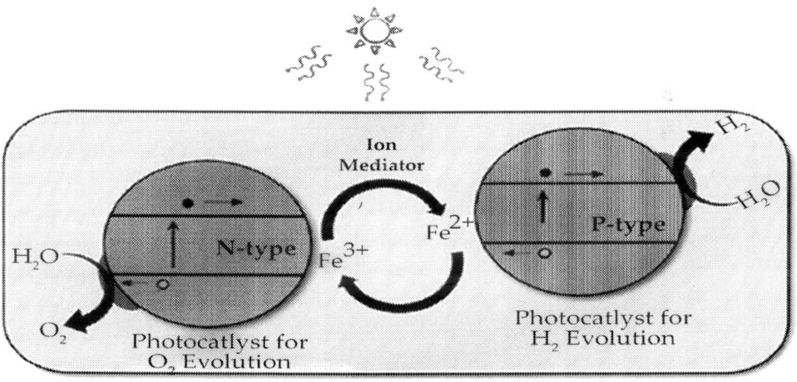

Figure 10. Energy band diagram of the Z-scheme showing OER and HER photocatalyst particles with functionalized surface sites for the oxygen and hydrogen evolution reactions, respectively. Concentration of both particle types are suspended together in solution with an ion-shuttling mediator, such as the Fe^{3+}/ Fe^{2+} couple, effectively coupling the gas evolution reactions in tandem to photosplit water. From (van de Krol & Gratzel, 2012)

PEC/PV devices

Due to the advances in single and multijunction solar cells, their combination with PEC devices can yield relatively high STH efficiency as a stand-alone device. In this case, the additional voltage that a photoelectrode needs to work independently is provided by a PV device, which would work with the transmitted photoelectrode light. (Ager, et al., 2015) Morisaki et al. realized this concept by combining a TiO_2 photoanode with a crystalline silicon solar cell. (Morisaki, et al., 1976)

Since then, these kind of devices have been significantly improved. (Abdi, et al., 2013) published a study combining a $BiVO_4$ photoanode with a Co based catalyst combined and a multijunction amorphous silicon based solar cell. The structure is illustrated in Figure 11. The final efficiency that was achieved was 4.9%.

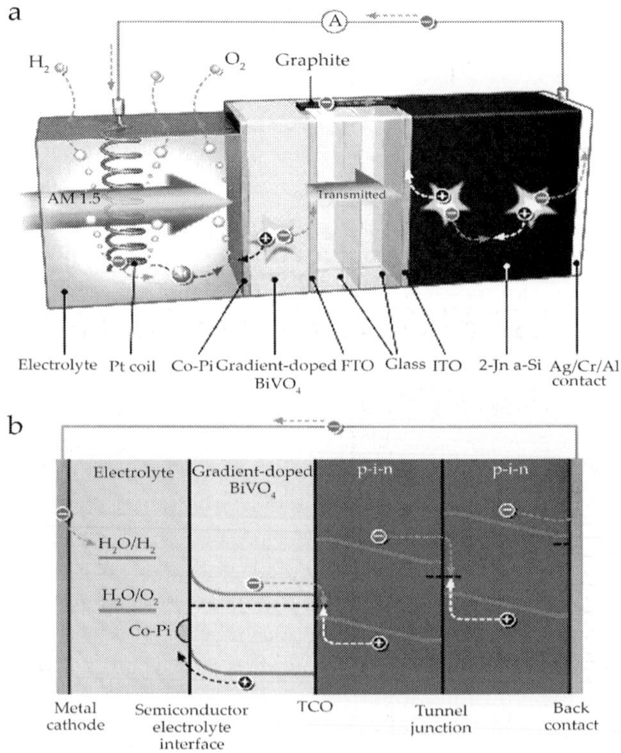

Figure 11. a) Schematic diagram of the combined W:$BiVO_4$ gradient doped and a-Si solar cell device. b) The corresponding band diagram of the hybrid device. From (Abdi, et al., 2013)

(Han, et al., 2015) extended this concept to a monolithic multijunction device in which the last junction relies in a semiconductor/electrolyte junction for charge separation. A Pt catalyst was added to the surface. This concept is illustrated in Figure 12. This device is able to achieve 5.1 mA cm^{-2} at 0 V vs RHE. It must be noted that this case shows a photocathode, while the previous device is based on a photoanode, and thus the connections and light direction would vary between these two devices.

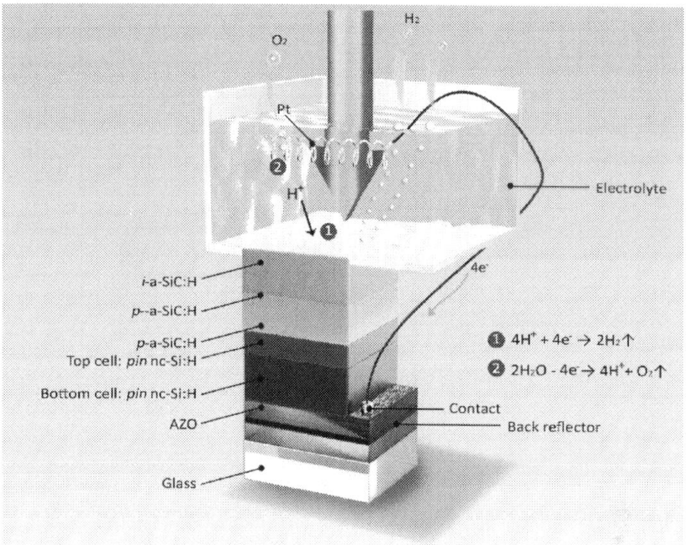

Figure 12. Monolithic aSiC/ncSi/ncSi PEC/PV device. From (Han, et al., 2015)

Miller et al. reported a combination of a WO_3 photoanode with a multijunction amorphous silicon based solar cell, which they are planning to upscale to a semi-industrial level. (Miller & Rocheleau, 2002) They determined that three junctions were needed to achieve the desired voltage levels.

So far, only earth-abundant materials have been studied. However, with other, more sophisticated materials such as p-$GaInP_2$ and a p-n GaAs photovoltaic cell, higher efficiencies can be reached, up to 12.4 %. (Khaselev & Turner, 1998) These devices would also use a Pt catalyst. Nevertheless, although these devices are more efficient, it is probably not cost efficient for large-scale devices.

A summary of the main reported combinations of PEC and PV devices is reported in Table 10. From this research, there are some general trends that can be envisioned. Firstly, most of the high efficiency devices use noble metals as catalyst, which is expected. Secondly, most of the solar cells used are either based on amorphous silicon multijunctions or III-V technologies. This is probably due to the flexibility that these kind of devices give in terms of the voltage and current provided, which would be defined by the different junctions and materials used.

Table 2. Summary of the most efficient PEC/PV devices so far. Adapted from (Ager, et al., 2015) and (Walter, et al., 2010)

Device structure	Conditions	STH efficiency	Stability	Reference
c-Si (pn) / WO_3 (n)	1 M H_2SO_4 1080 mW/cm_2 (11 suns)	0.0068 %	10 min, stable current	(Shaner, et al., 2014)
c-Si (pn) / Fe_2O_3 (n)	1 M Na_3PO_4 60 mW/cm^2	6.0 %	40 min	(Wang, et al., 2014)
c-Si (pn) / TiO_2 (n)	0.5 M H_2SO_4, 1.5 suns	0.12 %	4.5 h	(Liu, et al., 2013)
a-Si:H (pin) / $BiVO_4$ (n)	pH 7.3, 1 sun	3.6 %	1 h, stable current	(Abdi, et al., 2013)
DSSC / WO_3 (n)	1 M $HClO_4$ (pH 0) 1 sun	3.1 %	8 h, 30 % current drop	(Brillet, et al., 2012)
DSSC / Fe_2O_3 (n)	1 M NaOH (pH 13.6) 1 sun	1.17 %	8 h, 30 % current drop	(Brillet, et al., 2012)
$GaInP_2$ (p) / GaAs (pn)	3 M H_2SO_4, 11 suns	12.4 %	20 h, 20% drop in current	(Khaselev & Turner, 1998)
a-Si:H (pin) / nc-Si:H (pin) / $BiVO_4$ (n)	pH 7, 1 sun	5.2 %	1 h, less than 5 % current loss	(Han, et al., 2014)
a-Si:H (pin) / a-Si:H (pin) / WO_3 (n)	0.33 M H_3PO_4, 1 sun	3 %	Not reported	(Gaillard, et al., 2010)
DSSC / DSSC / Fe_2O_3 (n)	1 M NaOH (pH 13.6), 1 sun	1.36	Not reported	(Brillet, et al., 2010)
a-SiC (pi) / a-Si:H (pin) / a-Si:H (pin)	pH 2	1.6	Not reported	(Miller, et al., 2005)

Multijunction PV/electrolysis Approaches

Instead of using a photoelectrochemical device directly, the generation of charge carriers and redox reaction can be decoupled by combining a PV device with an electrolyzer or a direct PV device, eliminating the need of the electrolyte for charge collection and electric field generation. Thus, many of the devices reported in this section would not be completely decouple, since no electronic devices to adapt the voltage and current would be considered, and thus the voltage created by the PV device should still be able to drive the water splitting reaction. Thus, multijunction devices would probably still be preferred. (Ager, et al., 2015) collected all the most successful integrated devices from the last years, divided according to the number of junctions, the materials used and the arrangements regarding wires and catalysts, as shown in Figure 13.

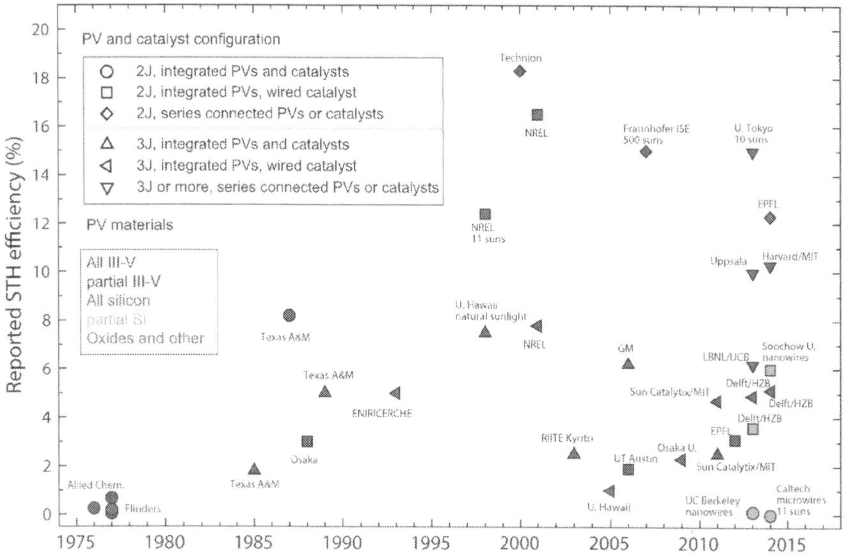

Figure 13. Reported solar to hydrogen (STH) conversion efficiencies as a function of year and sorted by the number of tandem photovoltaic junctions used (2 or 3). The fill color represents the semiconductor materials used in the photovoltaic portion of the device. All STH conversion efficiencies are as reported in the original publications. From (Ager, et al., 2015)

The first device considered really efficient was developed by NREL in 1998 (Khaselev & Turner, 1998), with the structure depicted in Figure. Among more recently and most noticeable devices is the III-V based cell from Technion in Israel, which yielded an efficiency of 18 %. (Licht, et al., 2001) This device consisted of a AlGaAs/Si RuO_2/Pt device, and accurately matched the potentials and band alignments needed for water splitting. Other devices based on more abundant materials include device fabricated in EPFL in 2014, with more than 12% efficiency. The device consisted of a perovskite solar cell, and used Earth-abundant catalysts. (Luo, et al., 2014) Unfortunately, the stability of perovskites needs further development, especially in a water environment. Thus, more stable devices based on oxides and silicon might also be considered as a viable option. (Schuttaufa, et al., 2016) connected three solar cells in series to achieve the required voltage for the reaction. A different approach was taken by (Urbain, et al., 2016), developing a quadruple junction amorphous silicon based solar cell, to be able to combine with a Ni catalyst. They were finally able to achieve an efficiency of 7.8 %.

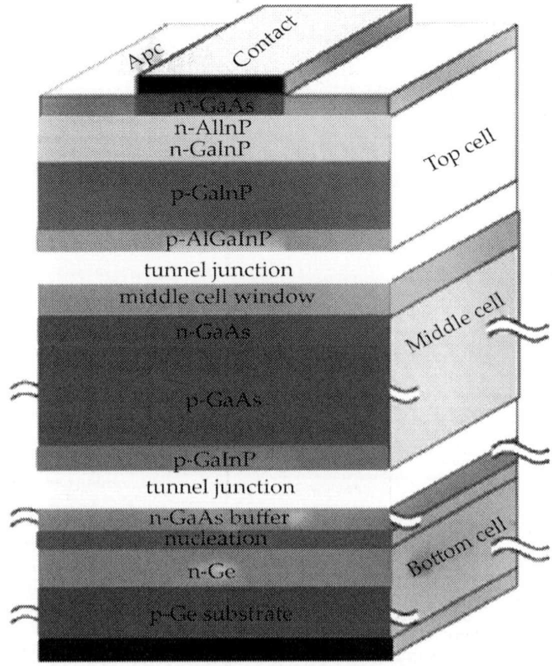

Figure 14. Device structure of a high efficiency $GaInP_2$/GaAs/Ge triple junction solar cell developed at NREL. From (Ager, et al., 2015)

Some authors combine the use of a multijunction device with the change of light direction, which might allow light scattering and light division among the different semiconductors. In addition, this strategy allows for light concentration, which would reduce the use of semiconductor, making III-V material based devices considered as a feasible approach. However, these kind of devices have the added challenge of defining the area, which would make the definition of STH efficiency ambiguous. (Ager, et al., 2015) The most efficient direct PV devices reported so far are summarized in.

Table 3. Summary of the most efficient direct PV devices. Adapted from (Ager, et al., 2015)

Device structure	Conditions	STH efficiency	Stability	Reference
Halide perovskite ($CH_3NH_3PbI_3$) 2 cells side-by-side	1 M NaOH 1 sun	12.3 %	2 h, stable photocurrent 10 h, small degradation	63
$Ga_{0.83}In_{0.1}As$ (pn) / $Ga_{0.35}In_{0.65}P$ (pn)	PH 7, 500 suns	15 %	2.3 h, stable current	84
$GaInP_2$ (pn)/ GaAs (pn)	2 M KOH, 1 sun	16.5 %	9 h, stable current	50
$Al_{0.15}Ga_{0.85}As$ (pn)/ Si(pn)	1 M $HClO_4$, 135 mW/cm^2	18.3 %	14 h stable current	(Licht, et al., 2001)
a-Si (pin)/a-Si(pin)	0.5 M H_2SO_4, 1 sun	2.93 %	Not reported	48

The trade-offs established when improving these multijunction devices are the loss on overall current density production, and the additional complexity and cost of the structure. To further optimize these device to compensate these disadvantages, a factor to take into account when making monolithic multijunction devices is the possible losses at the interfaces due to defects, and the tunneling recombination junctions must be optimized by increasing doping concentrations. Moreover, they must be current matched to minimize the losses. This can also be achieved by including light trapping strategies, such as middle reflectors, texturing or plasmonic particles. Finally, the possible lattice matching among the different crystalline materials used must be accounted for. If these challenges can be efficiently tackled, the efficiency of these devices could get closer to the theoretical limit of 46 % STH efficiency. (van de Krol & Gratzel, 2012)

The other possibility is to completely detach the two factors by including a PV system into an electrolyzer. Although these systems would include additional losses due to the cabling and power coupling, they would gain in flexibility of design, improving the separate components, which may lead to better performances, as shown in.

Table 4: Summary of the most efficient devices based on PV and electrolyzes. Adapted from (Ager, et al., 2015)

Device structure	Conditions	STH efficiency	Stability	References
Si minimodules with NiMoZn/NiB	pH 9.2, 1 sun	9.7 %	168 h, constant current	(Cox, et al., 2014)
GaInP$_2$ (pn)/GaAs (pn)/Ge (pn) with Pt/Pt	pH 9.2, 1 sun	6.2 %	15 h, constant current	(Modestino, et al., 2014)
CuIn$_x$Ga$_{1-x}$Se$_2$ (pn) 3 cells side by side, Pt/Pt	3 M H_2SO_4, 1 sun	10 %	27 h, stable H_2 production	(Jacobsson, et al., 2013)
GaInP(pn)/ InGaAs (pn)/ Ge (pn) with Pt/Pt	pH 7, 10 suns	15 %	1 h, constant current	(Fuji, et al., 2013)
a-Si (pin)/ a-Si (pin)/ a-Si (pin) with NiMoZn/Co	pH 9.2, 1 sun	4.7 %	24 h, 20% drop	(Reece, et al., 2011)
DSSC/DSSC/DSSC wired side by side Pt/Pt	2 M KOH, 1 sun	3.7 %	20 h, stable current	(Park & Bard, 2005)
a-Si (pin)/ a-Si (pin)/ a-Si (pin) with CoMo/NiFeO	pH 13, 1 sun	2.5 %	18 h, 10% drop	(Yamada, et al., 2003)
a-Si (pin)/ a-Si (pin)/ a-Si (pin) with Pt/Pt	2 M KOH, 1 sun	7.8 %	Not reported	(Khaselev, et al., 2001)
a-Si (pin)/ a-Si (pin)/ nc-Si (pin) / nc-Si (pin) with Pt/RuO$_2$	0.1 M KOH, 1 sun	9.5 %	4 h, stable current	(Urbain, et al., 2016)
a-Si (pin)/ a-Si (pin)// nc-Si (pin) / nc-Si (pin) with Pt/Ni	0.1 M KOH, 1 sun	7.5 %	4 h, stable current	(Urbain, et al., 2016)

MEMBRANES

Recently, a new trend has started of using a membrane to separate the solution between anode and cathode. This allows to maintain different pH in the two sides, which as previously explained would improve the performance of each half reaction. Moreover, having a membrane separating the two sides has further advantages, such as the easier separation of the two gas products, hydrogen and oxygen.

A cell showing the conventional configuration including a membrane is shown in Figure 15. In addition, the sketch of the band diagram of the system at short-circuit conditions is depicted. The membrane must be fabricated so that it allows certain ion permeation, but does not allow the gases pass and recombine. Thus, the membrane, often made out of a polymer, separates the photoanode from the Pt counter electrode. In addition, to allow water feed and gas escape from the cell, a segmented sub-gasket is used between the membrane and the electrodes, whereas water is injected into the cell through a hole drilled on the backside of the counter electrode.

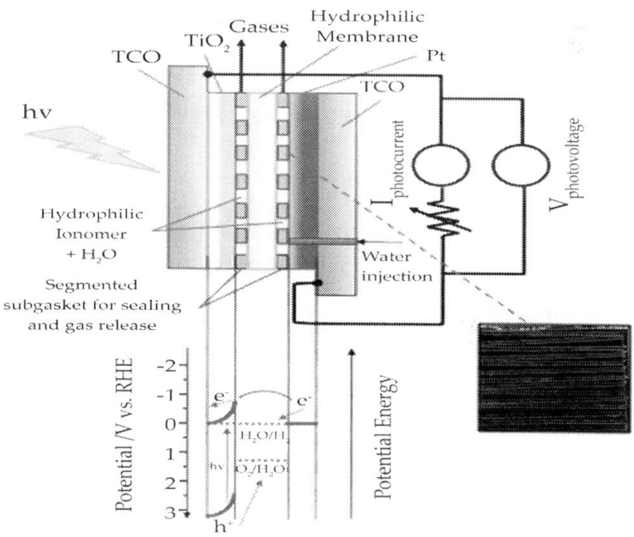

Figure 15. Sketch of the photoelectrolysis cell and the energy diagram for the water splitting process under the short-circuit conditions. Note that the Fermi levels are equal in the semiconductor and electrolyte. From (Arico, et al., 2017)

The principles of operation of these membranes can be better understood when looking at the band diagram shown in Figure 15. Solar irradiation would be generated and collected in the PEC device. When these charge carrier reach the semiconductor/electrolyte interface, they are used to drive the OER and HER in the anode and cathode, respectively.

(Arico, et al., 2017) By separating the conditions of these two reactions, better performances can be achieved in each of the two sides. In particular, the membrane/ionomer chemistry produced a different operating pH and this would influence significantly the photoelectrolytic conversion. Ultimately, the membranes can be tailored to modulate electronic levels at the interface in order to achieve the best properties to match with the semiconductor band gap and band edges. This might be especially useful for lower bandgap semiconductor materials, since they are able to harvest a larger fraction of the solar radiation. (Arico, et al., 2017)

The other factor to take into account regarding membranes is the separation between hydrogen and oxygen for easier separation, which would lower the cost of the overall PEC system. (Chen, et al., 2016) Gas crossover can affect the composition of both head spaces, which dilutes the product streams and even has the potential to create explosive gas mixtures. (Berger, et al., 2014) Thus, membranes to separate these two streams would be essential for good functioning of a practical PEC device.

The designs need to be able to transport enough ions without transporting too many gas molecules. Thus, these materials must have a high ionic conductivity and low gas permeability. Looking at these requirements, block copolymers are particularly attractive for obtaining the desired transport properties. These properties can be estimated by measuring the power curves and current-voltage curves. However, if there is substantial gas crossover the current density would increase, but the hydrogen production would decrease due to water synthesis, which shows that the STH efficiency should be calculated based on the hydrogen measure rather than the current. (Berger, et al., 2014)

One more property that must be accounted for is the transparency of the membrane, depending on the systems, since light might need to be transmitted through to arrive to a photoelectrode. That would depend both on the material and thickness of the membrane. Moreover, the amount of crossover and permeation would also depend on the thickness, and thus it is an important parameter to study, ranging in the nanometer scale. (Berger, et al., 2014)

The most useful membranes to be used in solar water splitting are probably bipolar membranes (BPM). The objective of these membranes is to decouple the conditions of each electrode solution. In order to achieve high efficiencies in a PEC device, strong bases or acids are needed in each electrode to increase conductivity and minimize concentration gradients. (Hernandez-Pagan, et al., 2012) However, earth-abundant based OER catalysts are not stable in acid, and the ones for HER are not stable in bases. This limits either the applicability of earth-abundant materials or the possibilities of the electrolyte optimization. By using BPMs, the pH incompatibility is decoupled, opening a new door to PEC device optimization. (Luo, et al., 2016)

During operation, the BPM blocks the anion transport from the cathode compartment to the anode compartment through the cation-exchange layer and the cation transport from the anode compartment to the cathode compartment through the anion-exchange layer. At the same time, water is able to penetrate through both layers of the BPM. Due to the built-in electric field, water starts to dissociate at the interface of the cation-exchange layer and anion-exchange layer, releasing both protons and hydroxyl ions. When an electric field is applied between anode and cathodes, these ions migrate through the BPM to the appropriate compartment. These processes help maintaining the desired pH at each compartment. (Luo, et al., 2016)

The materials used to fabricate these membranes are often based on polymers, since they are flexible, durable and efficient. Perfluorosulfonic acid (Nafion® 115) and quaternary ammonium-based anionic (Fumatech® FAA3) ion exchange membranes are two examples of these polymers. It must also be taken into account that there are anionic and cathodic membranes, which enhanced the performance of the semiconductors, since anionic membranes would increase the concentration of hydroxides with enhanced adsorption of OH species that promote the OER. (Arico, et al., 2017)

These membranes have been proved in a functional solar water splitting device by (Luo, et al., 2016). They used two perovskite based solar cell connected in series combined with a BPM for electrolyte and gas separation, compartmentalizing the electrolytes of 0.5 M H_2SO_4 and 1 M KOH, respectively. That is to allow the catalysts used, CoP and NiFe in the cathode and anode respectively at their optimum conditions so that it can reach better performances. This system is illustrated in Figure 16, achieving a STH initial conversion efficiency of 12.7 %. The conditions of this device consists in a pH13.9 for the anode side and 0.4 for the cathode side, which translates to a voltage at the BPM of 0.798 V, and it slightly increases with increasing current due to resistances inherent in the electrolyte.

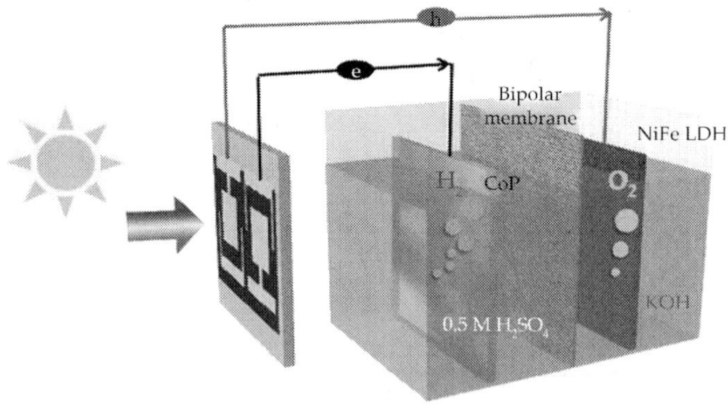

Figure 16. Schematic diagram of the solar-driven water-splitting device composed of perovskite light harvesters, Earth-abundant catalysts, and bi-polar membrane. From (Luo, et al., 2016)

Thus, this device proves the applicability of BPMs for water splitting, allowing each catalyst to work at their best pH, without posing big problems in their stability after 100 h. In addition, it maintains the two produced gases separated, to lower the purification costs and the chance of hazardous and explosive situations. (Luo, et al., 2016)

However, it must be taken into account that these systems should be scalable to be economically feasible. In particular, the BPM materials need to be cheap enough for large-scale applications by using Earth abundant polymers such as Nafion. However, the authors of these device report the perovskite cells to be the limiting factor in terms of applicability. (Luo, et al., 2016)

SEAWATER AS ELECTROLYTE

Another option concerning the electrolyte is the use of seawater. Seawater is widely available, it generally needs to be purified for other uses, and the salts naturally make the solution more conductive. (Mase, et al., 2016) In particular, Cl^- ions are oxidized more easily than water, and thus hydrogen can be produced without the need of decomposing the H_2O molecule. (Luo, et al., 2011) That would be integrated in the sustainable future vision, which can be applied in a wider variety of locations, shown in Figure 17. (Tachibana, et al., 2012)

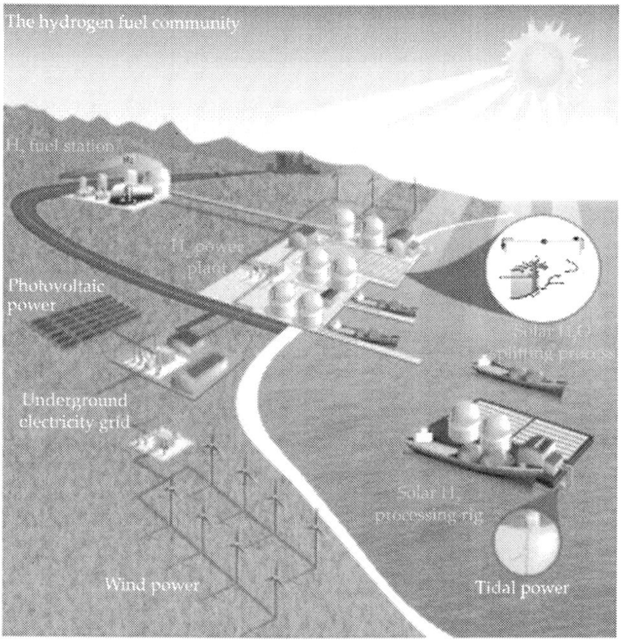

Figure 17. Vision of a sustainable hydrogen fuel community based on APS. Hydrogen is produced on a solar water-splitting power plant using seawater on floating ports. From (Tachibana, et al., 2012)

Interesting studies regarding the use of seawater, (Li, et al., 2015) used a TiO_2 with C_3N_4 and CoPi to sensitize the system both in the case of light absorption and chemical activity. This system was identified to be stable after 10 h of testing, with a current density produced of 1.64 mA/cm^2 at 1.23 V vs. RHE. Thus, this report shows that seawater does not cause additional degradation in the proposed system. Another study proposed $BiVO_4$ as photoanode for the same reason. This resulted in a photocurrent density of 2.16 mA/cm^2 at 1 V vs RHE, and being stable for several hours. (Luo, et al., 2011)

Thus, even though limited reports on the mechanisms exist up to date, the use of seawater for photoelectrochemical hydrogen production seems like a viable option. Although the reported efficiencies are lower than the reported using more traditional electrolytes, the stabilities are good and the costs can be reduced, and the concept of solar energy and water splitting can be applied in a wider variety of locations and applications.

SUMMARY

So far, the main principles and main materials for PEC devices have been explored. This section has gone further with these concepts, looking for new and emerging trends to improve the device efficiency. Here, the main trends explored were the use of nanotechnologies, the combination of several junctions and the use of membranes was explored.

Nanotechnologies can yield several advantages. First, the electrical properties can be improved, since by including structures such as nanopores or nanotubes the distance that charge carriers have to travel is reduced. Moreover, they can improve the light trapping and absorption capabilities by creating quantum confinement of photons. In particular, quantum dots are able to accurately tune the bandgap by changing the particle diameters. Examples of these different structures have been proven useful in TiO_2 in a wide range of literature reports.

Multijunctions can also improve the efficiency of PEC devices by combining several semiconductors with different bandgaps, which would increase the spectral utilization and the obtained voltage. There has been many efforts both from the PEC and PV fields, to increase the efficiency of multijunction devices over the past several decades. These efforts have yielded demonstrator with STH efficiencies as high as 18 %, and a calculated theoretical limit of about 46 %. These highly efficient devices, however, tend to use expensive materials such as III-V technologies, and can include very complex and costly configurations. Therefore, at the moment these integrated approaches have not evolved enough to be economically attractive. Despite these challenges, there seems to be a consensus among the field that in the near future, multijunctions are unavoidable in the route to high STH efficiencies and stand-alone devices.

The main structures identified were the combination of two photoelectrodes, in which they would still depend on the electrolyte for charge collection with respect to the band alignment; the combination of a photoelectrode with a solar cell that provide the additional voltage; and the use of a PV device directly or indirectly in combination with an electrochemical catalyst. In all these devices, the most important parameters to have into account are the bandgap combinations and number of junctions, as well as how they combine with the catalyst and electrolyte.

Connected with the catalyst and electrolyte conditions, another emerging trend is the use of membranes to separate the two electrode sections. That way, the pH can be tuned to very acidic or basic environments to increase the conductivity of the solution without damaging the catalyst. These membranes are generally made out of polymers, such as Nafion. These have been used together with a perovskite cell and earth-abundant based catalysts, achieving impressive efficiencies.

Finally, another method regarding the electrolyte regulation, seawater can be used instead of the distilled water that is normally used. This increases the conductivity, and the economic effectiveness of the system, since seawater is much more available than distilled water.

By applying these new emerging techniques to PEC devices, their efficiencies could be highly improved, and the costs could be reduced due to the use of the earth-abundant materials. However, these technologies have only been proven in small scale. Thus, they should be proven into a larger scale before they can be considered economically feasible. In addition, the fabrication of these devices is important, and thus, that would be considered in the next section.

CHAPTER 5

ELECTRODE FABRICATION

As previously mentioned, the method used to fabricate the photoelectrodes can have a great impact on the performance of the final device, both in terms of STH efficiency and stability. Thus, here the main fabrication methods will be explained, together with their advantages and disadvantages.

SUBSTRATE CONSIDERATIONS

It is important when preparing and measuring an electrode to expose only the area of the material of interest, rather than other materials involved such as the metal contacts. In addition, proper substrate preparation would ensure a good ohmic contact between the substrate and semiconductor, reducing resistances and increasing the performance of the photoelectrode. Finally, the fabrication method, crystal structure of the target material and substrate type would affect the final performance of the device, and thus must be taken into careful consideration.

An ohmic contact must be ensured to ensure optimal charge carrier transfer, since it does not rectify the electric current or inject minority carriers into the semiconductor bulk. However, ohmic contacts present a certain resistance, which must be studied and minimized. These contacts are often using transparent conductive oxides (TCOs) or metals. It is important to note that not only the resistivity of these layers must be considered, but also their work function and band alignment with respect to the semiconductor. Finally, some contacts may alloy with the semiconductor, resulting in different materials, which could have beneficial or detrimental consequences for the electrode performance. Materials such as flour-doped tin oxide (FTO) stainless steel, aluminum, titanium or molybdenum are valid candidates as substrates for these kind of devices. However, some of these materials are not compatible with deposition processes such as metal-organic chemical vapor deposition (MOCVD), and thus have limited applicability. A list of substrates and their compatibility with certain techniques and materials are summarized in Table 1. (Chen, et al., 2013).

Table 1. List of substrates compatible with some semiconductors. From (Chen, et al., 2013)

Semiconductor	Deposition process	Known compatible substrates	Reference
CGSe	Evaporation	FTO	(Marsen, et al., 2008)
WO_3	Sputtering	FTO, ITO	(Marsen, et al., 2008) (Gaikwad, et al., 2005)
$GaInP_2$	MOCVD	GaAs	(Khaselev & Turner, 1998)
GaPN	MOCVD	GaP, Si	(Deutsch, et al., 2006)
InGaN, $GaInP_2$	MBE	SiC, sapphire	(Khaselev & Turner, 1998)
Fe_2O_3	Electrodeposition + calcination, sputtering	FTO, glass/Ti/Pt	(Kay, et al., 2006) (Kleiman-Shwarsctein, et al., 2008)
Cu_2O	Electrodeposition	Ti, ITO, FTO	(Siripala, et al., 2003) (Nian, et al., 2008)
a-SiC	PECVD	ZnO_2, FTO, stainless steel, Cr	(Matulionis, et al., 2008)

SEMICONDUCTOR DEPOSITIONS

Some used deposition techniques have already been mentioned in the previous section. Thus, there are many possible techniques to deposit such materials. However, each of these techniques has advantages and disadvantages that should be considered when choosing a fabrication method. The ideal method would be able to produce highly uniform and conformal materials, with the desired crystallinity and a high control over the layer thickness. The most sophisticated methods, which lead to the higher quality materials, tend to be expensive, use high vacuum system and high temperature treatments, which might not be feasible when applied to an industrial scale. The most prevalent example is atomic layer deposition (ALD) a self-limiting technique that is able to grow very conformal monolayers, but is extremely slow, and thus does not hold great industrial promise. In any case, in order to better understand the advantages and disadvantages of each of these techniques, they should first be briefly described. The techniques considered here to be the most commonly used are , molecular beam epitaxy (MBE), chemical vapor deposition (CVD), and electrode position. (Chen, et al., 2013)

Physical Vapor Deposition (PVD)

These techniques include evaporation and sputtering processes. They are performed in vacuum chambers and use a solid source of the target material. The material is then either heated (evaporation) by a thermal source or an e-beam, or etched using a plasma (sputtering). Once these processes take place, the vaporized target materials would reach the substrate and condense there. There is a wide variety of materials that can be deposited using these techniques, from metal oxides to nitrogen compounds or metals. They tend to have reasonably good quality and can be controlled quite precisely, but they need high vacuum and are considered to be relatively expensive for mass-production processing. In order to better understand which parameters would affect the material quality the most, each method would be looked into in more detail.

Evaporation

In general, the aim of this technique is to heat the material source, which includes mostly a metallic compound, above the melting point, reaching the evaporation of the material. Then, the evaporated material would condensate on a cooler surface, which in this case would be the substrate. This process is conducted in a high-vacuum chamber, in order to control the thickness and oxidation of the deposited layer. The homogeneity of the layer can be further improved if the substrate is rotating throughout the deposition. (Seshan, 2002)

Evaporation can be achieved by two methods: resistive evaporation or electron beam evaporation. In the case of resistive evaporation, the material source is placed in a crucible, which is heated resistively by applying a high current. On the other hand, for electron beam evaporation, the source is placed in a cooled crucible, and it is irradiated by an intense electron beam in order to heat the source and reach evaporation. The electron beam is achieved using a tungsten filament directed by strong magnetic fields. (Seshan, 2002) Both techniques are illustrated in Figure 1.

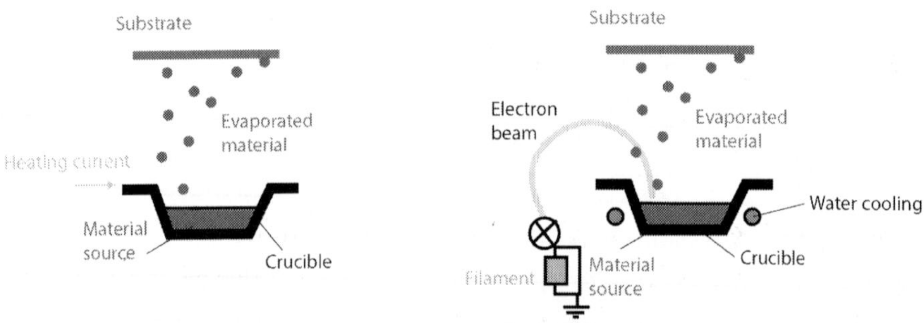

Figure 1. Schematic representation of evaporation techniques, including thermal evaporation (left) and e-beam evaporation (right)

These techniques have the limitation of the metal contained in the crucible and the low deposition rate. Thus, they are usually applied when a thin, conformal layer of the material is needed. However, it cannot be used commercially to create thick layers of material due to their low throughput and the need for high vacuum equipment.

Another consideration that must be mentioned, especially in the case of e-beam evaporation, is the fact that, if the crucible is not rotated or the e-beam is not moved around the crucible, there is a possibility of creating a hole through the material and into the crucible itself, causing unwanted effects such as the deposition of the metals of the crucible of damaging the water source. Thus, even though e-beam evaporation tends to be more effective and can achieve higher rates, thermal evaporation appears as a less hazardous option. In addition, thermal evaporation can be applied to all kinds of materials, while e-beam needs of a specific dielectric constant to be able to evaporate the material.

Sputtering

Even though other techniques such as CVD can achieve higher purity and better surface roughness and film adhesion, this technique can achieve similar characteristics with much lower costs. (Anon., 1999) This method is most commonly used for the deposition of metals and metal oxides. In particular, it is often used in the electronics and solar cell fields to deposit transparent conductive oxides (TCO) such as ZnO, SnO_2 or InO_3, but can also be used for other materials such as metals like Pt or Ni.

In this method, an accelerated atom or ion transmits its energy and momentum to the source atoms placed in a target via sequential collisions. Thus, the target atoms can escape and move to the gas phase. The accelerated atoms are achieved using a power supply called magnetron, either using reactive direct current (DC) or radio frequency (RF). DC sputtering can only be applied to conductive materials, whereas RF sputtering can also be used for dielectric materials. Once the target atoms have reached the gas phase, they would deposit on the substrate (anode). The deposition chamber is kept at a certain pressure by an inert gas such as argon (Ar), which would be accelerated by the oscillating electromagnetic field, which would be the ones to bombard the target surface. The schematic representation of this method is shown in Figure 2. If nitrogen is used instead of argon, nitridic coatings would be achieved. (Smets, et al., 2016)

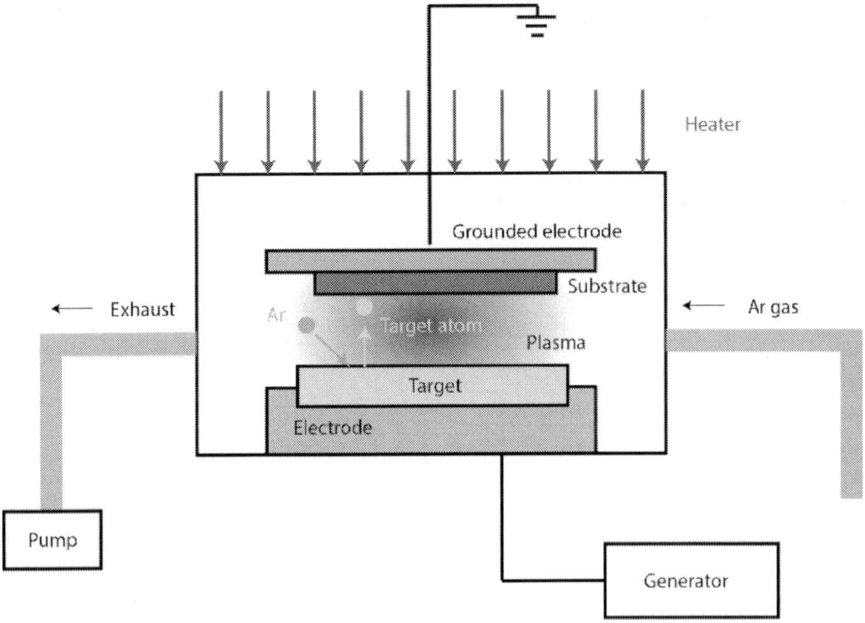

Figure 2. Schematic representation of a sputtering setup.

As already mentioned, this method results in rather good layers with a high throughput. However, it must be noted that the resulting layers have a higher defect density than the layer resulting from CVD methods. In addition, the use of Ar to displace the target atoms creates ions, which can cause ion bombardment and subsequently damage the substrates surface. In order to minimize this effect, processing conditions such as pressure, temperature and RF power should be carefully controlled. (Seshan, 2002)

Molecular Beam Epitaxy (MBE)

This could be considered a PVD process in some ways. The main difference here is that the control of the vapor flow in MBE allows the growth of monocrystalline films, as opposed to the amorphous and highly disordered films that PVD techniques tend to yield. To achieve these high crystallinities, this method requires a crystalline substrate that would act as a seed. This method needs slightly less demanding deposition conditions that PVD while still yielding good material properties, and thus could be considered more economically feasible. However, the need for a crystalline substrate increases its cost, and thus a trade-off is still prevalent.

Epitaxial growth refers to the growth of a crystal layer on a single crystal substrate. (Fisher, et al., 2012) It can be used to either grow a layer within a device, or a whole wafer which can be separated from the original wafer. The growth occurs by using different source gases, such as silane, dichlorosilane, thrichlorosilane, diborane, phosphine or arsine among others. Depending on the gases uses, also dopants can be introduced in the structure. The chambers are sealed, using a hydrogen ambient. Due to the stable conditions maintained on the chamber, it has a relatively good uniformity control. This method thus lies within the chemical vapor deposition methods, which will be further explained later on in more detail. (Schneider, et al., 1987)

However, this method can cause localized stresses appearing in this method can cause cracks and crack propagation. The crack growth can be controlled by temperature tuning, since the wafers also suffer from thermal stresses due to the lattice mismatch between the substrate and the deposited material. (Kajari-Schroder, et al., 2013) If the bottom wafer is removed from the substrate, the latter can be used again to grow other epi wafers for several cycles. (Schneider, et al., 1987) This allows reducing the use of high quality material, since the substrates can be reused and there is no loss of the material due to cutting and dicing.

Among the advantages is also the fact that relatively large wafers can be produced. In addition, the energy consumption is significantly lower than for other methods. Looking at the disadvantages of this method, the necessary high temperatures to reach the desired materials is one of the limiting factors. In addition, this is mainly a batch system due to the gases and purging needed, which makes the process more expensive. There are some defects that can appear within this method, including lattice dislocations, impurity particles from the surroundings, inhomogeneities in the thickness of the layers, with marked hills within it, and stacking faults due to incorrect nucleation or due to faults in the substrate. (Schneider, et al., 1987)

Chemical Vapor Deposition (CVD)

This family of techniques include Metal-organic chemical vapor deposition (MOCVD), low-pressure chemical vapor deposition (LPCVD), plasma-enhanced chemical vapor deposition (PECVD) and atomic layer deposition (ALD), among others. These techniques tend to yield the best material qualities and performances. They are based on the decomposition of one or more volatile molecules, which act as chemical precursors. These molecules would decompose or react on the substrate enhanced by the different sources of energy (plasma, pressure, chemical potential, etc.). Once they decompose, they would reattach to the substrate surface, creating the semiconductor layer. They are highly controllable processes, leading to single crystal, polycrystalline or amorphous materials depending on the deposition conditions. However, they also demand a high amount of energy for the process to occur, and they are often very slow, making them hard to implement in an industrial process. Thus, even though they lead to the best quality and most conformal layers, they are also often considered to be confined to lab scale devices due to their high costs.

Plasma Enhanced Chemical Vapor Deposition (PECVD)

In this technique, a flow of a mixture of gases is introduced in the very high-vacuum deposition chamber. Then, by applying a certain electrical power within two electrodes inside the chamber, a plasma is created. The conversion is regulated by the electric field created in between the two electrodes, with an radiofrequency (RF) signal of 13.56 MHz or by a very high frequency (VHF) signal at ~50 MHz. The plasma produces reactive radicals and ions from the gases in the chambers, due to the high-energy collisions in the gas phase, which would interact with the substrate and deposit a thin film of these free radicals and atoms. The quality of the layer is controlled by the temperature and pressure control, as well as the gas flows and the power used to create the plasma. (Smets, et al., 2016) The process is illustrated in Figure 3.

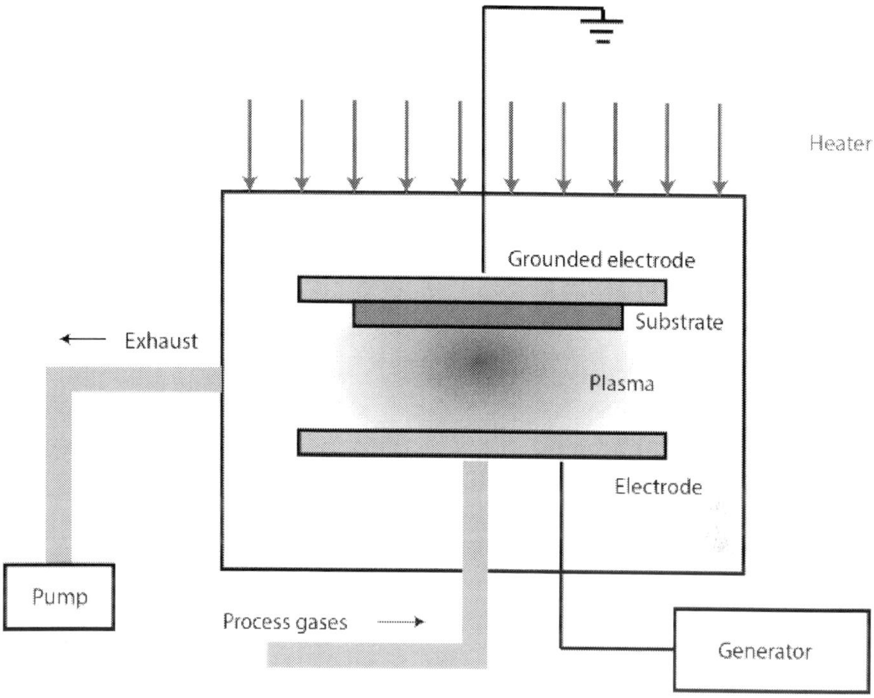

Figure 3. Schematic of the plasma-enhanced chemical vapor deposition (PECVD) system

Thin film materials such as amorphous silicon, nanocrystalline silicon and their alloys such as silicon carbide or silicon oxide can be deposited by PECVD. In addition, by adding the right gases, they can be n or p doped. Typical gases used for these materials are silane (SiH_4), hydrogen (H_2), methane (CH_4) or carbon dioxide (CO_2). In addition, borane (B_2H_6) and phosphine (PH_3) are traditionally used to include the doping elements for p and n-type doping, respectively. (Smets, et al., 2016) This method can also be used to grow other materials as carbon nanotubes (Meyyappan, et al., 2003) or monomers (Vasudev, et al., 2013)

The technique offers the advantage to be able to deposit uniform and homogeneous layers with good adhesion to a wide variety of substrates, both flat and textured. In addition, the chambers need to be kept at temperatures as low as 200°C. However, the high vacuum characteristics limit its applicability. In addition, if crystals are desired, this method is not optimal. The resulting layer can be crystallized with further processing, such as the ones described in the previous section. Another factor to take into account is the volatility and hazardousness of the precursor gases. (Jones & Hitchman, 2009)

Metalorganic Chemical Vapor Deposition (MOCVD)

This method allows the growth of high quality multilayer structures with good uniformity regarding both thickness and composition. It has often been used as an epitaxial growth method for electronic and photonic device processing. However, these layers can also be used as thin films on a certain substrate. This technique allows for the deposition of new device concepts even at large scale. It is especially popular as a way to deposit epitaxial III-V materials such as InP, GaAs or InSb.

In this technique, metalorganic and hydride precursors are used in gas phase. Usually, alkalis of the group II and III metals, and hydrides of the group V and VI are used as precursors. The choice of precursor generally aims at improving the layer properties, achieving low deposition temperatures, reducing premature reactions and improving the safety of the process. The overall mechanism of this method is illustrated in Figure 4. The main requirements for a good precursor are sufficient volatility for acceptable growth rates, adequate stability to assure decomposition ant the desired rate without premature reactions, free from residual impurities, and not to create secondary reactions in the gas phase. Inadequate choice of precursors can lead to carbon contamination, as it is the case in MOCVD deposition of AlGaAs due to incomplete elimination of methyl groups from [Al-CH_2] species. Another possible problem is the concentration control of the precursor in the gas phase, especially in ternary and quaternary compounds, which would result in lattice mismatches and poor uniformity. (Holloway & McGuire, 1995)

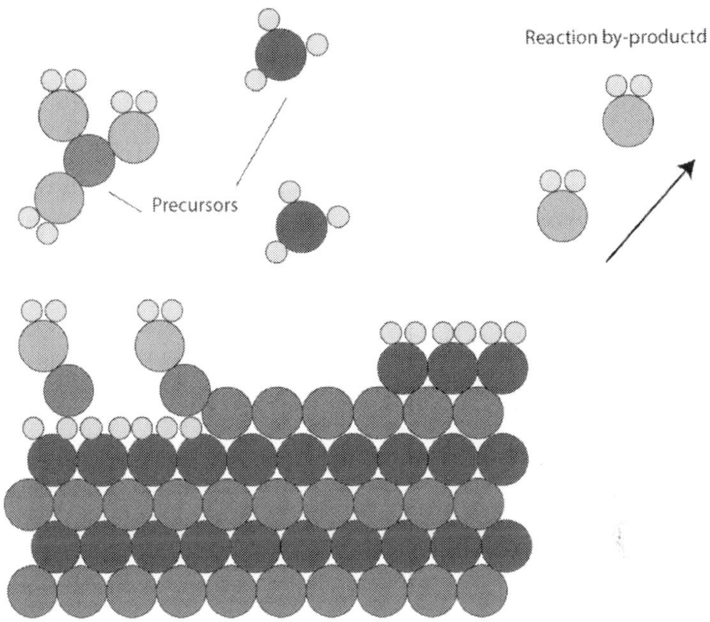

Figure 4. Schematic of the mechanism involved in the MOCVD process

During the process, a thermodynamic driving force supersaturates the gas phase. The growth precursors are then introduction into the reaction zone, and their concentration must be high enough to ensure enough decomposition and supersaturation. Thus, the stability of the precursors is an important parameter in the process.

Once the gases reach the reaction zone, there are four main steps:

- Mass transport of reactants to the growth surface or substrates
- Reaction on the substrate surface
- Incorporation of the new material to the substrate surface
- Removal of the by-products that resulted from the reaction at the surface.

The slowest step from this four would be the rate limiting step of the deposition. *For example*, for GaAs and InP, it has been found that the mass transport of the group III precursor is limiting the overall process.

In must be highlighted again that one of the main disadvantages of this method is the security issues associated with the precursors traditionally used, which are usually highly toxic and flammable such as amine and phosphine. Solutions to this problem include the use of precursors based on liquid alkyls, which are approximately one hundred times less hazardous; or the generation of the hydrides in-situ either from solid precursors combined with hydrogen plasma or by electrochemical methods. In addition, a safe disposal must also be considered. Dangerous metalorganic residues are often oxidized in order to neutralize them. Moreover, in dry scrubbing systems, active carbon and other appropriate materials are used to neutralize these compounds.

Atomic Layer Deposition (ALD)

Also known as atomic layer epitaxy (ALE), this technique targets the need for smaller devices that need more spatially demanding structures. It has been used to deposit a wide range of materials, including metals, semiconductors and insulators, both amorphous and crystalline. The most common materials grown by ALD are oxides, nitrides, sulfides and single elements. However, a much wider variety is available, only limited by the availability and volatility of particular precursors. These materials have successfully been used in the microelectronic and solar cell industry as dielectrics and buffer layers. (Johnson, et al., 2014)

In this technique, two precursor gases are sequentially introduced in the deposition chamber in the form of pulses. The gas introduce in each pulse would react with the substrate surface. The individual reactions of each gas are also called "half-reactions," since they only provide part of the final material. Once the first gas has been introduced for the duration of the pulse, in which it should have fully reacted with the substrate surface, it is evacuated and purged using an inert gas, and the second gas (counter-reactant) is introduced to lead the same process. It is a cyclic process, in which the amount of cycles determines the thickness of the layer. Each cycle is a self-limiting reaction that leaves a monolayer on the surface. The overall process is illustrated in Figure 5. (Johnson, et al., 2014)

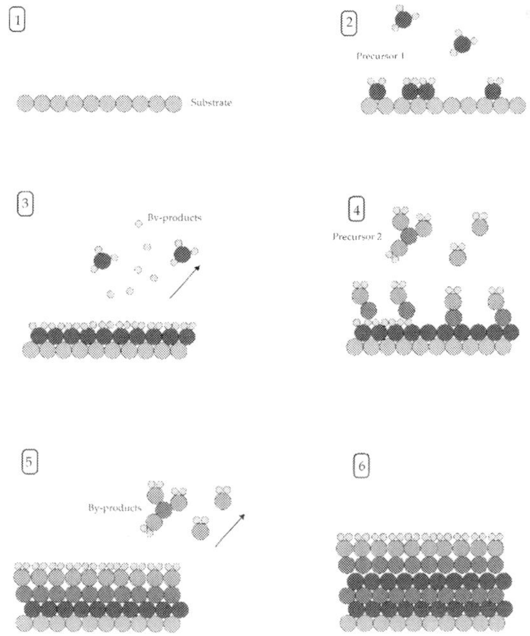

Figure 5. Schematic of the ALD mechanism, where step 1 is the initial clean substrates, step 2 is the pulse of the first precursor, step 3 is the purge of the by-products and unused precursor, step 4 is the introduction of the second precursor, step 5 is the purge of the by-products and unused second precursor, and step 6 is the finalized film.

This process needs in general less extreme conditions in terms of temperature and pressure than the CVD techniques, typically less than 350 °C and not necessarily high vacuum conditions. The temperature range where the growth is saturated is called the "ALD temperature window", and depends on the specific process and material to be deposited. This is represented in Figure 6. If temperatures outside this window are used, other reactions might happen, leading to non-ALD depositions due to precursor condensation (low temperature) or thermal decomposition (high temperature). (Johnson, et al., 2014)

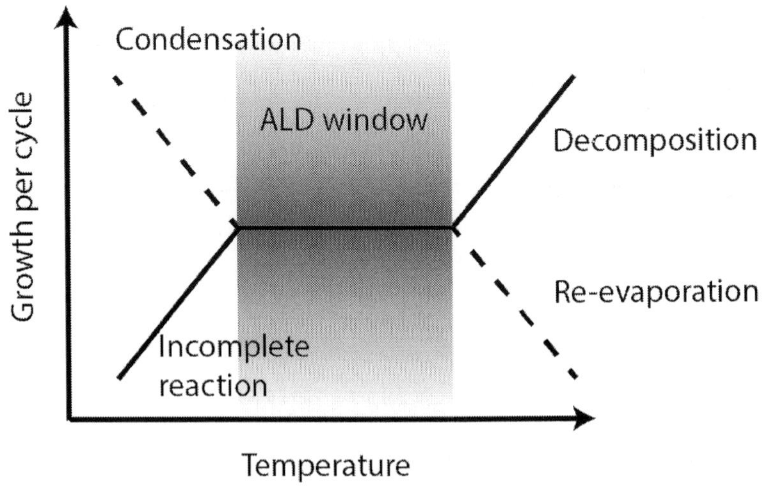

Figure 6. Schematic representation of the ALD temperature window

The advantages of this deposition method over alternatives are the conformality and extremely high control over the thickness and composition of the material deposited by tuning the cycle rates and the number of cycles. However, regarding the disadvantages of the method, the non-linearity of the relationship between cycle ratios and atomic ratios complicate ternary and quaternary compound depositions. In addition, those process also need to take into account the individual temperature windows of each of the precursors, which might not be compatible. Unfortunately, choosing reactants always involves a trade-off between reactivity, volatility, safety, availability and cost.

Finally, the deposition rates are very slow compared to other techniques, in the order of 100-300 nm/h. (Johnson, et al., 2014) Moreover, as the substrate area increases, the deposition rate decreases due to the longer pulses needed to saturate the surface. In addition, it must be noted that the deposited films are often amorphous, and to achieve a certain crystallinity that would improve the electrical properties often necessary, the material needs to be annealed, increasing the time consumed and costs of the process. To overcome these shortcomings, spatial ALD shows promising results.

Electrodeposition

This technique is based on the electrochemical reaction between a conductive substrate and an electrolyte solution that contains a salt of the target material. When an electrical potential is applied between the substrate and the counter electrode, an electrochemical reaction will take place at the surface of the substrate, resulting on a layer of the reacted species. Thus, the anode would be consumed in the process, while the metal ions would be deposited on the surface of the cathode. (Raub, 1982) The schematic of this setup, together with the reactions that occur in the electrodes, is shown in Figure 7.

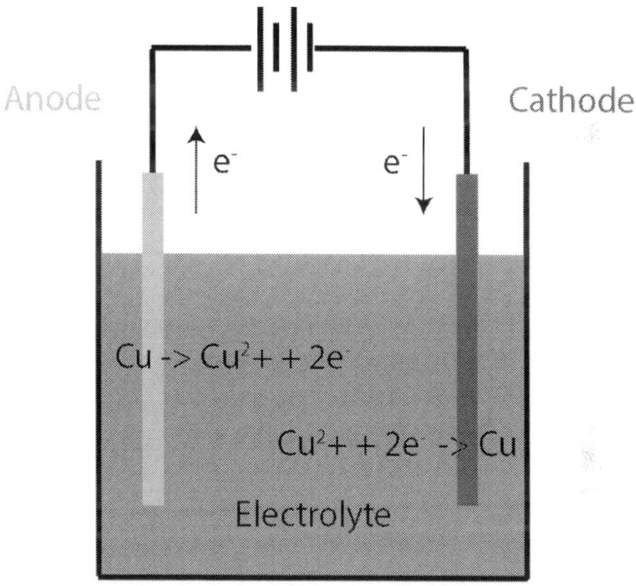

Figure 7. Schematic of the electroplating system. Note that in this case the deposited material is copper in the cathode. The electrolyte contains a solution of $CuSO_4$

This process can be controlled by the current and voltage applied by the power source. In addition, the number of valence electrons in the metal and the temperature would determine the deposition rate. Mathematically, the ideal case can be expressed as

$m = ItA_wzF$ Equation 38

where m is the deposited mass, I is the electric current produced, t is the time that the process took, Aw is the atomic weight, F is the Faraday constant (96485.3365 As/mol) and z is the number of electrons transferred per deposited atom.

This method is simple and cost-effective, can be carried out at ambient conditions, and produces high quality layers, which makes is very attractive for large scale applications. Although this is the simplest and more economically feasible of all techniques presented so far, it is also worth noticing the limitations of this method. It is restricted to the deposition of electrically conductive materials on electrically conductive substrates, producing polycrystalline and amorphous materials of lower quality than the methods previously discussed. (Raub, 1982) (Raval & S.Solanki, 2013)

CATALYST DEPOSITION

As previously discussed, the semiconductor properties play an important role on the PEC process. However, catalysts are also essential to extract the charge carriers at the semiconductor/electrolyte interface to drive the redox reactions. Thus, in order to have high quality catalysts, their deposition techniques must also be considered. Note that the deposition technique would also depend on the selected catalyst, which varies depending on the nature of the reaction. Some of the most common catalysts for the HER are Pd, Pt, Ni and Co. However, in the case of the OER, more appropriate catalysts are RuO_2, IrO_2, and mixture of Ni and Co with metal oxides. (Chen, et al., 2013)

Regarding to the properties of the layer, conformal layers are often desirable, in order to also act as a protective layer. However, the thickness should not surpass that of a few monolayers due to several reasons. First, the catalyst should be transparent to avoid shadowing. In addition, the metal can act as a concentration of recombination centers, since most metals are able to collect both electrons and holes. To avoid this, a discontinuous coverage of catalyst is recommended, with an interparticulate distance smaller than the diffusion length of the charge carriers. In order to create these structures, the deposition method is crucial. (Chen, et al., 2013) Thus, the main deposition methods must be discussed here, including electrodeposition, sputtering and spray pyrolysis.

Electrodeposition

As previously explained, they are simple and low-cost techniques based on electrochemical principles. It is possible to deposit conformal layers, but also catalyst islands, controlling the dimensions of the islands by adjusting parameters such as the applied potential, current density, or concentration of the precursor. Periodic pulsed profiles can also yield different morphologies by adjusting the frequency and intensity of the pulses. Finally, if performed under illumination, catalyst electrodeposition can result on orientations favorable to the water splitting reaction, which would further enhance the performance. (Chen, et al., 2013)

Sputtering

PVD processes in general yield conformal films with minimum thicknesses around 100 nm. However, it is also possible to deposit material islands if the process is stopped in the early stages. In that case, small islands will be present, before the material coalesce into a conformal layer. Such process needs low deposition rates and very controlled environments. One more variable to consider is the fact that sputtering techniques also use the rotation of the substrates to increase the homogeneity of the films. However, in the case of catalyst deposition, since conformity of the film is not desired, it is recommended not to rotate the sample. (Chen, et al., 2013)

Spray Pyrolysis

This method is best suited to deposit metal oxides such as RuO_2, and it is limited to semiconductors that remain stable at relatively high temperatures. In this technique, the substrate is placed below a nozzle at a certain distance. Then, a precursor solution containing the ions of the target compound is sprayed on it using a carrier gas, which is often inert. This spraying is done in cycles, and the substrate is heated in order to evaporate the solvent solution and precipitate the target compound. The schematic representation of the setup can be seen in Figure 8. The number of cycles, time of each cycle, temperature, nozzle distance and precursor solution are the main parameters to be controlled in order to achieve the desired compound with the target crystallinity and thickness. In some cases, a subsequent annealing step would be needed to achieve the final crystallinity of the compound, and possibly decrease the amount of defects in the material.

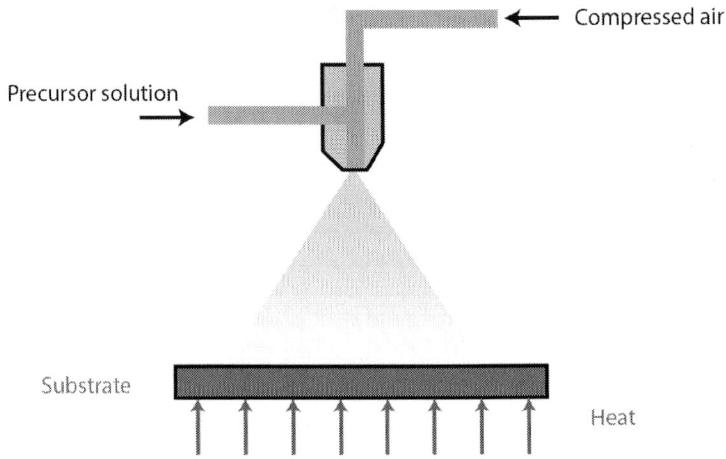

Figure 8. Schematic of the spray pyrolysis setup

The advantages of this method, as stated before, are their simplicity, low investment cost and relatively mild conditions. In addition, it could be implemented in a roll-to-roll process. However, the resulting layer might not be of a comparable quality to more sophisticated methods such as CVD techniques, and might need additional steps such as annealing to achieve the desired properties. Moreover, to achieve higher quality material slower cycles are needed and, due to the cyclic nature of the method, it can be rather time consuming, with a relatively low throughput (Chen, et al., 2013)

ELECTRODE CONNECTIONS

When depositing a photoelectrode by using the techniques previously described, some more practical considerations must also be taken into account. It is important that these electrodes can also provide a good way to be connected to the external circuit. These connections must comply with certain requirements: large area to create an ohmic contact, electrical and chemical isolation of all conductive hardware and from the electrolyte. Thus, different mounting structures can be considered. (Chen, et al., 2013)

Insulated Electrical Wires

In order to achieve a uniform potential distribution and enhance charge carrier collection, the wires must be correctly chosen, taking into account their material resistivity, cross sectional area and length in contact with the electrode substrate. Most of the wire will be covered by an insulating coating that assure separation from other components, and only one portion is stripped to be soldered to the edge of an uncovered part of the substrate. The soldering material must be considered to assure high conductivity, such as indium or silver. Once the contact has been established, all uncovered parts must be insulated by using an epoxy resin, to minimize the appearance of pinholes. T must be considered that the epoxy resins must be curated in order for them to harden and provide effective insulation, but the curation process must be performed in a way that it does not damage the electrode. (Chen, et al., 2013)

Copper Tape and O-ring

This method is used for cases when multiple samples are arranged in a library array, with the test cell moving to measure one ample after another, or when epoxy encapsulation is not possible. Then, an O-ring is used to seal the contact between electrode and electrolyte. This ring should be coated with a material inert in the electrolyte, like Viton or Teflon, to avoid contamination. Another advantage of this method is that further characterization after PEC testing is possible, which would otherwise not be the case if the epoxy resin was present. (Chen, et al., 2013)

ELECTRODE SURFACE AREA DETERMINATION

The performance of a PEC electrode must be normalized to the electrode surface area that was active during characterization. Thus, it is important to determine the planar illuminated area. Two main approaches can be considered.

Digital Picture

By counting the number of pixels on a digital picture of the electrode, the active area can be determined. Several software packages can aid on this method when the scale of the picture is clearly shown. The accuracy of this method depends then both on the quality of the picture and the processing software used. (Chen, et al., 2013)

O-Ring

As previously mentioned, an O-Ring can be used for insulating the electrode connections from the electrolyte. An additional advantage is that it is able to accurately define the area in contact with the electrolyte without the need of further image processing or other complicated methods. However, it is important to notice that the compression of the O-Ring may lead to possible errors at the edges, and thus by using a bigger O-Ring these errors can be minimized. (Chen, et al., 2013)

SUMMARY

Several techniques can be used to fabricate the different parts of the devices, such as PVD, CVD, ALD, or electrodeposition. Previous sections have discussed some of the most appropriate techniques to deposit the different materials. Here, a more in depth analysis of the processes and parameters have been carried out.

First, it is important to consider the kind of substrate and cleaning process that would be used, since this can determine the amount of impurities and defects that would appear in the system. In addition, for this PEC application is also important to look at their durability and band effects of the substrate and TCO when integrated in the device.

When looking at the semiconductor and catalyst deposition, there are many different methods to deposit these materials, and depending on them the quality and cost of the process would be affected. In general, more expensive and slower methods yield better materials, but are less economically attractive, establishing a trade-off between these two factors. However, the method itself is not the only factor to account for. In addition to that, the parameters such as pressure, deposition rate, or power should also be taken into account, since they can drastically change the resulting material. Moreover, post-processing techniques such as etching or annealing could be used to further process these photoelectrodes and improve their characteristics.

Finally, once the main materials have been fabricated, the connection, mounting and characterization should also be taken into account. Here, two techniques to mount and connect these devices are presented: insulated wires and O-ring. Each of them have their advantages and disadvantages. Among the most important advantages of the O-ring is the well-defined area that shows, which eases the task of defining its efficiency. Note that these methods are mostly considered for lab scale, and further strategies might need to be developed for further application in practical devices.

CHAPTER 6

PEC CHARACTERIZATION

When designing and determining the performance of a PEC device, it is important to understand the physics behind the PEC process, and how the deposition techniques used may affect the performance of the photoelectrodes. However, understanding how to properly characterize these materials and devices is also a fundamental part of the process which is often overlooked. Proper characterization helps identifying potential issues, performances an allows a fair comparison between the different materials and device configurations. (Chen, et al., 2013) have described a method that comprises several characterization techniques in order to determine the performance of a PEC device and its suitability to be used at a bigger scale. The basic flowchart of this method is presented in Figure 1.

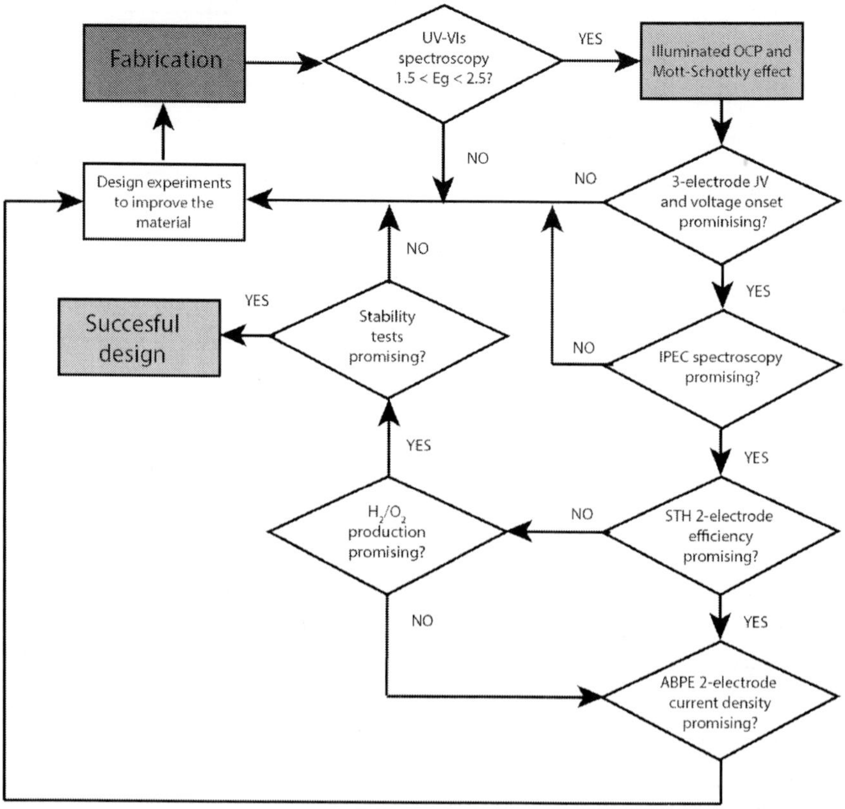

Figure 1. Recommended PEC characterization flow chart for a single absorber. From (Chen, et al., 2013)

From this flowchart, a list of the key characterization techniques used can be extracted, as shown in Table 1. In order to better understand the advantages and disadvantages of each of these techniques, they will be further analyzed in this section, dividing them in three main sections: optical, electrical and chemical characterization.

Table 1. Main characterization techniques, their output and disadvantages. From (Chen, et al., 2013)

Technology	Knowledge gain	Limitations
UV-Vis spectrum	• Optical bandgap • Direct or indirect bandgap • Limited transitions	• Non-conclusive information, prone to interpretation error • Good absorption does not directly translate into PEC efficiency • The bandgap determined is the one for the bulk, not necessarily the surface
Illuminated open-circuit voltage	• Flat-band potential • Conductivity type	• Possible photocorrosion at open circuit conditions • High recombination can avoid flattening of the bands
Mott-Schottky / Hall effect	• Flat-band potential • Conductivity type • Doping density • Density of charge carriers	• Complicated interpretation of data from non-ideal samples • Samples should be annealable and deposited on non-conductive substrates
Three-electrode cyclic voltammograms	• Saturated photocurrent density • Voltage range of photocurrent generation • Conductivity type • Flat-band potential	• Catalysts may be needed to minimize overpotential errors in flat-band potential measurement • Possible sample corrosion
IPCE	• Wavelength dependent incident photon to current efficiency • Bandgap • Absorbed photon to current efficiency	• High variability in data if the lamp source is not accurately measured • Possible photocorrosion reactions would lead to false higher IPCE
2-electrode zero-bias photocurrent	• Solar to hydrogen (STH) efficiency	• Possible photocorrosion or side reactions would lead to false higher STH efficiency

2-electrode applied bias photocurrent	• Applied bias photon-to-current efficiency (ABPE)	• Possible photocorrosion or side reactions, which will lead to false higher ABPE
H_2/O_2 gas detection	• Faradaic efficiency for water splitting	• Quantitative detection requires perfect sealing of the reaction vessel
Stability test	• Lifetime of material	• Instability of illumination source over time

OPTICAL CHARACTERIZATION

Optical characterization refers to the techniques that explore how a semiconductor electrode interacts with light, and how this light is able to create charge carriers. The two main techniques used for optical characterization in PEC devices are ultraviolet-visible (UV/Vis) spectroscopy, incident-photon-to-current efficiency (IPCE) and scanning electron microscopy (SEM).

UV-Vis Spectroscopy

In this measurement, light absorption is measured as a function of wavelength, and can also provide information about the band gap and electronic transitions of the material. The fraction of light transmitted in the material can be described as the Bee-Lambert law, which states that the fraction of light measured after interaction with the sample versus the incident intensity is dependent on the path length through the sample, the absorption cross section and the differences between energy levels. (Chen, et al., 2013)

The absorption patterns would depend on the energy transition and forbidden steps in the material. The absorption spectroscopy results is related to the Beer-Lambert law (Reuter, et al., 2015)

$$I(\lambda) = I_0(\lambda)e^{-k(\lambda)L} \qquad \text{Equation 1}$$

where $I(\lambda)$ is the transmitted radiation, I_0 is the incident radiation, L is the absorption path-length and $k(\lambda)$ is the wavelength dependent absorption coefficient. The absorbance can then be calculated as:

$$Absorbance(\lambda) = -ln\left(\frac{I}{I_0}\right)$$

Equation 2

By using this methodology, the bandgap of the semiconductor materials can also be determined. In addition, by the slope of the absorbance, the tail states could often associated with defects, allowing estimating the quality of the material. (Floriano, et al., 2010) A method to determine the bandgap of the material is by using Tauc plots, naming the value calculated as the Tauc gap. These graphs show the energy of light (hν) on the x-axis and the absorbance related variable, $(\alpha h\nu)^{1/r}$, in the y-axis, where α is the absorption coefficient. The exponent *r* represents the transition nature (direct or indirect bandgaps), where 3/2 is used for direct transitions and 3 is used for indirect transitions. Based on this conversion, a linear regime is observed. By extrapolating this line, the intersection with the x-axis would define the Tauc gap of the material. (Tauc, 1968) This graphical process is illustrated in Figure 2.

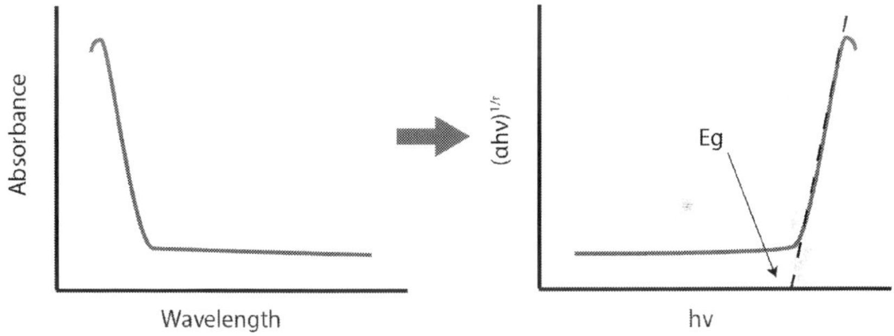

Figure 2. An example of an optical absorption spectra of a thin film, and the subsequent representation of the Tauc plot to determine the bandgap of the material

Thus, Uv-vis spectroscopy gives a convenient method to estimate the optical bandgap, which can also be used as an approximation for the electronic bandgap. In addition, it can also show the potential of different light trapping techniques, reflection patterns, etc. (Chen, et al., 2013)

The method is rather simple, with light hitting the sample at different wavelengths, and a sensor measuring either the reflected (R) or transmitted (T) light. The absorbance is then defined as 100 % - R – T, which essentially says that all the light that has not been reflected or transmitted must have been absorbed. One of the most common techniques to measure the absorption in a sample is the integrating sphere (IS). (Gaigalas, et al., 2009) In this system, a hollow sphere covered with a very reflective coating is used. There is a small aperture in the sphere, where the light enters and hits the sample, placed on the other side. The reflected and transmitted light will be measured by several detectors placed at different positions within the sphere. The role of the sphere is to reflect all light into the detectors, making sure that no light is "lost." A closer look and schematic representation on the IS is given in Figure 3. Once both transmission and reflection are known, the absorption of the material can be determined by the equation

$$Absorption\ (\lambda) = 100\% - Reflection(\lambda) - Transmission(\lambda) \quad \text{Equation 3}$$

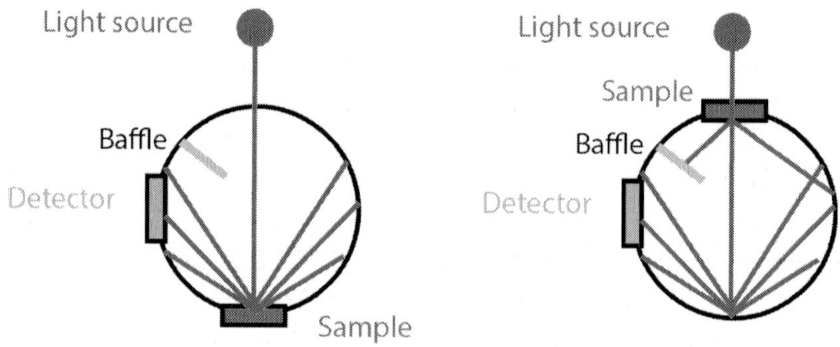

Figure 3. Schematic representation of the integrating sphere for reflectance (left) and transmittance measurements (right)

However, there are also some error sources in these calculations, since the assumption is that reflection and transmission only depend on the absorption characteristics, while in reality other interfaces and the medium itself might have an effect, especially if the sample is incorrectly placed or tilted, leading to non-zero baselines. To avoid some of these effects, lens filters can be added to the system. In addition, the determination of the bandgap zfrom these measurements is not straight forward. Different methods exist to determine it, being Tauc the most common one, and thus the obtained bandgap can be considered somewhat subjective. Finally, the light source would not have the same intensity at all wavelengths, and thus, it cannot provide the same accuracy in all points. It is especially difficult at low wavelengths, since most lamps used have a limited amount of UV light. Deuterium lamps have a range between 160-400 nm, while xenon lamps have a range of 200-2500 nm, and thus are usually preferred due to their wide range (Chen, et al., 2013)

Incident Photon-to-current Efficiency (IPCE)

This is a measure of the amount of charge carriers collected in the electrolyte per incident photon as a function of wavelength. Thus, this measurement takes into account the photon absorption efficiency, charge carrier excitation, separation efficiency, and transfer efficiency to the electrolyte.

Although in principle all this information is given as a combined feature, there are also methods to separate the different variables and efficiencies from each other. For example, by applying a voltage bias or combining these measurements with UV-Vis or EIS measurements, or by depositing a catalyst at the surface to ensure good charge carrier transport to the electrolyte, the individual components can be deconvoluted. In addition, this measurement can be used to determine the bandgap, since after that wavelength the IPCE would drop to zero. However, this is not as accurate as absorption measurements, since the onset might be limited by other factors such as slow kinetics or electron transport, giving falsely high bandgap values. (Chen, et al., 2013)

The system consists of a light source, a monochromator that adjusts the light wavelength, cut-off filters, a mechanical shutter, and connection cables and a potentiostat to measure the current produced. In addition, a secondary light can be added in case bias light is needed. The main elements setup is illustrated in Figure 4. The monochrome light is shined on the samples while at the same time the photocurrent is being measured with time.

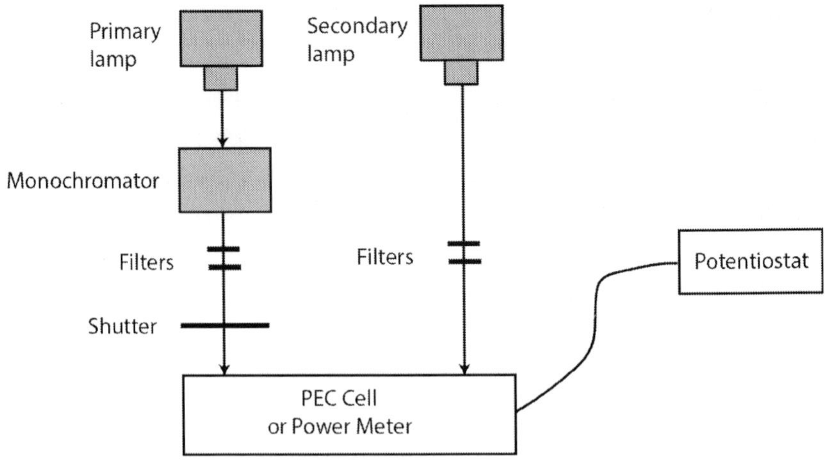

Figure 4: Schematic diagram of a typical experimental setup for measuring IPCE. From (Chen, et al., 2013)

The light source used would determine the wavelength range of this measurement, since it must emit photons at all measured wavelengths. It is also important that the intensity of the light source is high enough (100-1000 W) to achieve an acceptable signal-to-noise ratio. In addition, to avoid further inaccuracies related to the lamp intensity, calibration is needed every time the system is turned on. Finally, it is important for the area of incident light to be accurately determined, which is also taken into account during calibration. (Chen, et al., 2013)

Regarding the possible bias applied, an applied bias voltage should be performed either if the majority charge carrier is not able to drive its respective half reaction in a two-electrode measurement, or if an applied electric field is needed to aid charge carrier collection. If the performance highly differs when a bias voltage is applied, insight can be gained into the limitations of these two scenarios. Another case might be the need of a white light bias. This is necessary when the intensity of the monochromated light is insufficient to overcome the losses inside the materials. Thus, the additional background light enables more charge carriers to be generated, and thus assuring improved collection efficiency. In the case of using bias light, chopping the main monochromated is needed, to be able to differentiate between the effects of the white light and the monochromated light. (Chen, et al., 2013)

Another factor to take into account during these measurements is the chemical stability of the photoelectrodes during dark and illumination, which could change the results and make the interpretation of the data rather challenging. Thus, the focus in these cases should be on fining and electrolyte that minimizes corrosion, and take care to make the measurements as fast as possible. Finally, corrosion can be minimized by starting measuring from the long wavelengths to shorter wavelengths (from low to high expected response). (Chen, et al., 2013)

Scanning Electron Microscopy (SEM)

This is a well-established method to examine semiconductor materials and devices. This method allows geometric and morphologic characterization. (Heydenreich, 1993) It can provide information on the surface topography, crystallinity, chemical composition and electrical behavior. The deepness of this analysis goes up to about 1 μm. A semiconductor field is used to determine the different layers needed in a device, as well as the surface morphology and the and the edge morphology. The resolution of this technique can go up to 1000x.

In this method, a beam of electrons with energies between 2 and 40 keV are incident onto the sample. Then, several lenses demagnify the electron beam into a fine probe, which moves along the sample surface. The electrons penetrate on the semiconductor surface to a depth determined by the energy of the beam and the atomic mass of the ample material. The interactions between the electron beam and the material produces particles such as electrons, x-rays and luminescence, which are measured and translated into an image by a rasterization procedure. This equipment is schematically shown in Figure. It is important to notice that, if the semiconductor is deposited on an insulator such as glass, a conduction path to ground is required to avoid overcharging of the material. (Vernon-Parry, 2000)

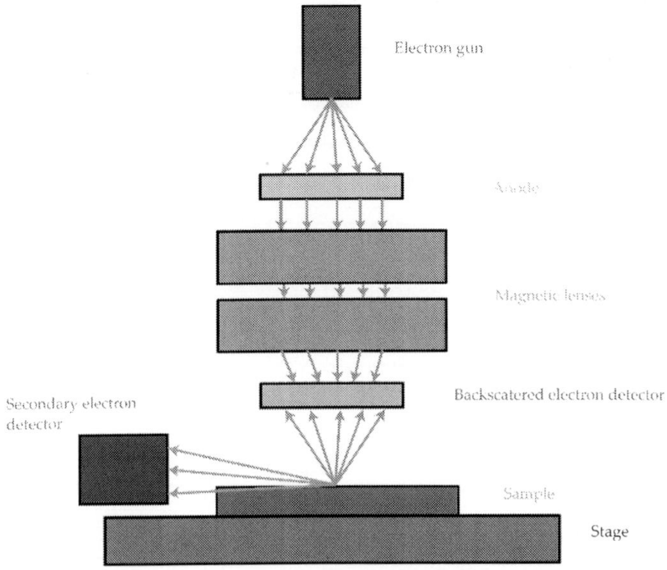

Figure 5. Schematic representation of an SEM setup and lens system

However, there are also some limitations. If the layer thickness is too small, the electrons accelerated on the sample might be transmitted through the sample. If the layers are thick enough, the electrons are no longer transmitted, and only particles (electrons, x-rays, etc.) emerging from the surface give information, which are recorded into conventional SEM images. (Vernon-Parry, 2000)

Several improvements can be included by changing the sample stages, to make them cold or hot, considering the effects of cathodoluminescence with temperature, where it is much stronger and less noisy at temperatures near absolute zero. In addition, magnifications up to 1,000,000x can be achieved by more advanced methods, which translate into a resolution of 1 nm. (Vernon-Parry, 2000) With that resolution, small features and defect atoms can be detected. (Fang, et al., 2015)

This technique can also be combined by other measurements such as cathodoluminescence (CL) to determine optical and electronic properties of these materials. The setup for this technique is shown in Figure 6, where the combination of lenses, fibers and electronics involved are depicted. In particular, this method can clarify the band structure and impurity levels, as well as doping levels (Yacobi & Holt, 1986)

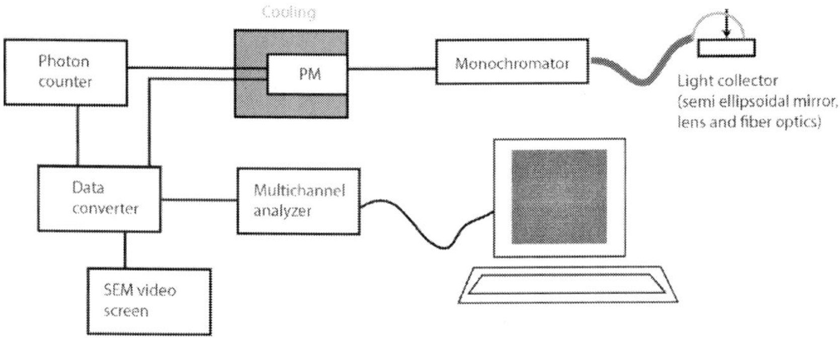

Figure 6. Schematic of the main components in a system combining SEM and CL detection for the visible range

ELECTRICAL CHARACTERIZATION

The PEC efficiency of these devices are closely related to the electrical performance of the semiconductors and their structure. These measurements mostly explores the conductivity, charge separation and collection characteristics of the device. Among the most interesting measurements are those of the open-circuit potential (OCP), electrochemical impedance spectroscopy (EIS), and JV characteristics. However, since many of these measurements are done under a solar simulator, and considering that this is an important source of inaccuracies, considerations over the used solar simulators would also be introduced.

Open-circuit potential (OCP)

This technique can be used to estimate the flatband potential if the illumination is high enough. In addition, it indicates the potential that the photoelectrode is able to provide to the reaction. The basic premise is to measure the potential difference created in the system at open-circuit conditions when in dark conditions and illuminated, as illustrated in Figure 7. That would result in the maximum photovoltage that can be achieved. However, it must be noted that the operational photovoltage would be slightly different, since at the operating conditions a photocurrent is produced, unlike in the case for open-circuit conditions. (Chen, et al., 2013)

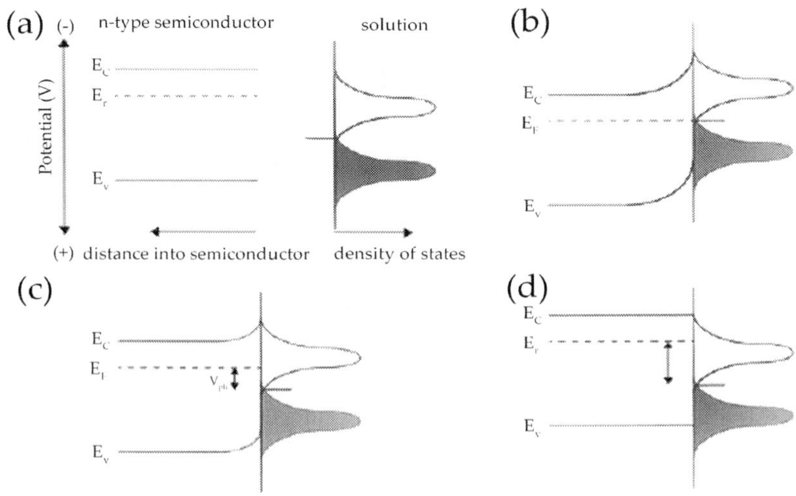

Figure 7. Band diagrams of a n-type semiconductor, a) prior to contact with the electrolyte solution, b) in contact with the solution in absence of illumination, c) in contact with the solution in the case of moderate illumination and d) in contact with the solution in the presence of intense illumination. From (Chen, et al., 2013)

The setup used to measure OCP uses a digital multimeter or potentiostat, which would often use a three electrode configuration. In addition, a solar simulator would be used to excite the sample during the light-on process. The light would then be turned on and off, while the potentiostat would register the voltage produced in the sample. It is important to note once again that the light intensity is a very important parameter, since the measurements would be inaccurate when the intensity is not high enough. A high enough illumination should saturate the sample, in a way that if more light is provided, the results would not change further, as illustrated in the example measurements in Figure 8. (Chen, et al., 2013)

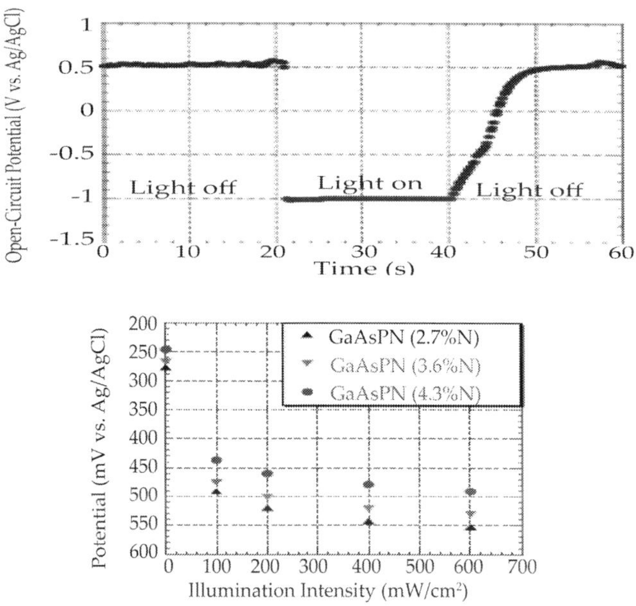

Figure 8. Open-circuit potential of a $GaInP_2$ electrode in pH 2 buffer. The lights are first off and then they are turned on, measuring the difference as the OPC vs Ag/AgCl. From (Chen, et al., 2013)

The conductivity type of the material can be determined using this analysis. If there is a positive shift (towards more anodic potentials) with illumination, the material is considered p-type. On the other hand, if the shift is negative (towards more cathodic potentials), the material would be n-type. (Chen, et al., 2013)

In addition, the OCP can be measured in a range or pH values, as shown in Figure 9. Most materials would follow a Nernstian relationship with pH due to the dynamic equilibrium at the surface, in which hydride/hydroxyl terminations tend to shift the band edges by -59 mV per pH unit. Thus, by considering the pH, the ability to move the band edges to a different energy level can be measured. Finally, considering forming structures such as p-n or p-i-n junctions can aid the charge separation and collection, improving the OCP values. (Chen, et al., 2013)

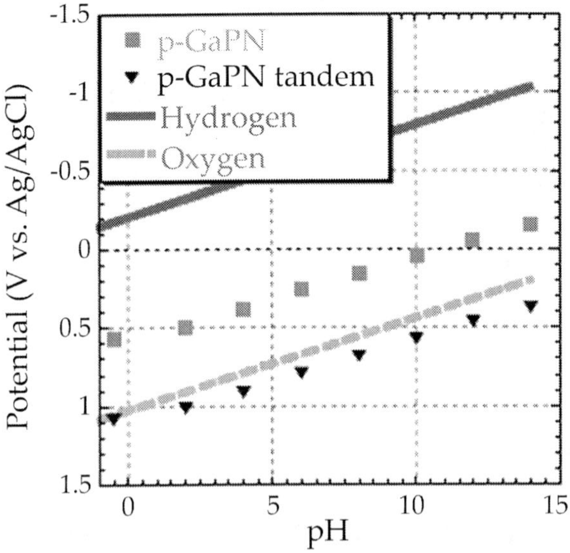

Figure 9. Potential vs. pH for p-GaPn and p-GaPN tandem. Both materials show Nernstian behavior. It is desired that the material characteristic is above the hydrogen line and below the oxygen line. From (Chen, et al., 2013)

One important factor to take into account is how the stability of the semiconductor might affect these measurements. To protect the photoelectrode from these conditions, it is recommended to reduce the exposure and timescale of the experiments to a few seconds or minutes. This short timescale would also reduce other undesired effects such as heating of the samples and solutions by the light source. (Chen, et al., 2013)

The results of this measurement give information regarding both the defects inside the semiconductor material and the effectiveness of charge separation. Low OCP represent high recombination rates, which might be related to the presence of numerous defects that act as recombination sites. (Chen, et al., 2013)

Electrochemical Impedance Spectroscopy (EIS)

This technique is used to measure the capacitances, impedances and resistances of the system, and can be made in dark and illuminated conditions. These measurements can then be fitted to an equivalent electric circuit, which would give information about where the main barriers for charge carriers are located.

Mott-Schottky plots

One of the most common fittings that can be applied are based on the Mott-Schottky (M-S) relationship, which involves measuring the capacitance of the space charge layer (C_{SC}) as a function of the applied potential (E), as (Chen, et al., 2013)

$$\frac{1}{C_{SC}^2} = \frac{2}{\varepsilon_r \varepsilon_0 A^2 e N_d}\left(E - E_{fb} - \frac{kT}{e}\right)$$

Equation 3

where ε_r is the relative permittivity of the semiconductor, ε_0 is the permittivity in vacuum, *A* is the surface area, *e* is the charge of an electron, N_d is the free charge carrier density, *k* is the Boltzmann constant, *T* is the temperature and *E* is the applied potential.

Based on this expression, both the flat-band potential (E_{fb}) and the free charge carrier density of either donors or acceptors (N_d) can be calculated, allowing for the determination of the band structure, along with bandgap information. These plots are depicted in Figure 10 .This will thus determine if the material has the right alignment with the water splitting potentials. In addition, the N_d also plays a role in the bulk and surface semiconductor properties, such as the width of the depletion layer and the recombination rate. Finally, the type of semiconductor (n or p) can also be revealed, with p-type materials having a negative slope and n-type materials a positive slope. c

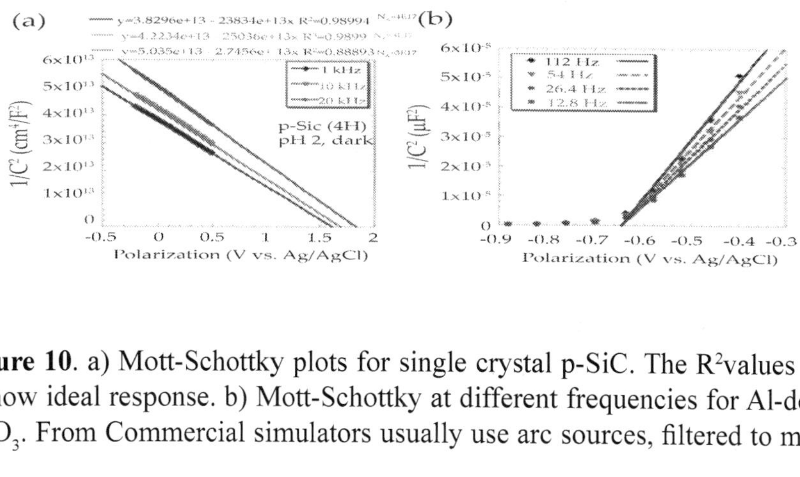

Figure 10. a) Mott-Schottky plots for single crystal p-SiC. The R^2values near 1 show ideal response. b) Mott-Schottky at different frequencies for Al-doped Fe_2O_3. From Commercial simulators usually use arc sources, filtered to mimic

the AM1.5 G spectrum, although they do not replicate it exactly. These simulators are usually calibrated to reach the short-circuit current equivalent at that under real AM1.5 G spectrum. That is usually done using reference cells that have been previously calibrated with a known spectrum. These cells can also be used to assure the stability of the solar simulator over time. To approximate this spectrum more closely, some solar simulators include two lamps, one of xenon and one of tungsten, since have a better spectrum at low wavelengths (Doscher, et al., 2016), while other are better at high wavelengths. That also assures that the simulator is valid at all bandgap semiconductors. (Chen, et al., 2013)

The equipment needed for this measurement is a potentiostat with a frequency analyzer and a three electrode setup. First, this experiment is done in the dark at a single frequency at a range of potentials to determine the voltage dependence of the space charge capacitance. This is then repeated for several frequencies, applying a reverse bias potential relative to the expected flatband potential. In addition, the potentials should be chosen to avoid any Faradaic reactions, in order to avoid charge transfer resistances. (Chen, et al., 2013)

It must be taken into account that M-S is a complicated method and analysis, since there are many parameter that might affect the measurements, such as the crystallinity of the sample, morphology or composition. Thus, a full evaluation of the frequency range and previous research fitting the data is helpful and must be explored. In particular, the frequency needs to be fast enough for the timescale to be small enough to allow for effective filling and unfilling of surface states of the build-up of a double-layer capacitance. (Chen, et al., 2013)

Bode and Nyquist plots

The Bode plots are able to determine individual charge transport processes, since small changes in impedance can be easily determined by this method. On the other hand, Nyquist plots, although they have a similar objective, they are less able to identify small impedances, and only the main processes are seen. Another difference of the two different plots is the frequency, where Bode specify it while Nyquist do not. Some examples of these plots and the different parameters that can be fit within them are depicted in Figure 11.

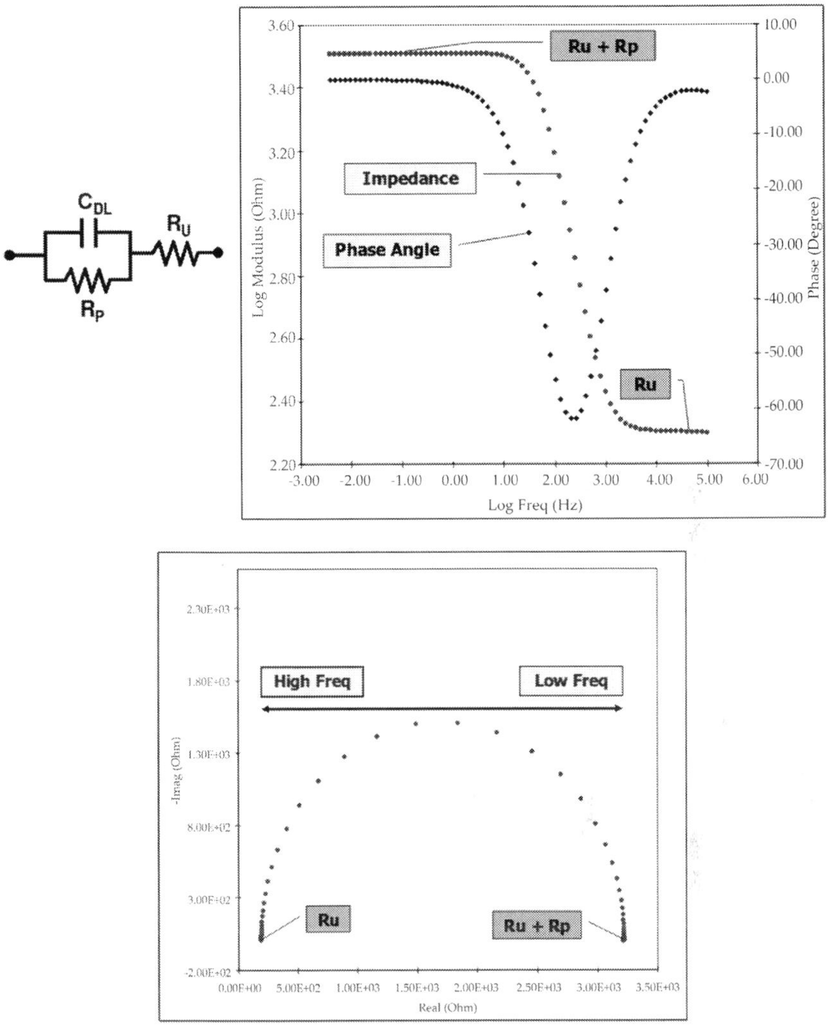

Figure 11. Bode plot (left) and Nyquist plot (right), including the equivalent circuit and the different areas of the plot. From (Gamry instruments, n.d.)

JV and Photocurrent Measurements

JV stands for current density – voltage characteristic. This is one of the most reliable ways to determine the potential STH efficiency of a photoanode. It consists of illuminating the material with light above its bandgap, creating charge carriers and a photocurrent. The measure consists of a voltage sweep, measuring the photocurrent produced at each voltage. The moment at which there starts to be photocurrent is called onset potential (E_{onset}), which is the offset relative to the flatband-potential by the overpotentials present. This voltage sweep can also be performed in dark conditions to obtain the so-called dark current. This would appear especially at high reverse bias due to the lowering of the Schottky barrier height and thus reverse bias breakdown of the semiconductor. In this region, starting at $E_{dark-onset}$, the surface of the semiconductor starts behaving as a metal. Thus, the characterization of the photoelectrode is most relevant within these two potentials. In must also be taken into account that this range may change with the pH, due to the changes in the bad potentials. (Chen, et al., 2013)

An important aspect to consider is if two or three electrode configurations are used. The general PEC system, consisting of the working electrode (WE) and the counter electrode (CE) is considered to be a two-electrode system, and gives a measure of how the overall system would act. If an additional reference electrode (RE) is introduced, a three-electrode configuration system is formed. This electrode is used as a reference to decouple the effects and limiting factors from the anode and the cathode, facilitating the diagnosis of any possible problems in each side. These two configurations are also illustrated in Figure 12. It must be noted that the counter electrode should be chosen based on the characteristics and band alignment of the working electrode for optimum performance. In addition, if the area of the counter electrode is increased, it is assured that the reaction is not limited by the kinetics of the counter electrode. Finally, the counter electrode must also be chosen. Althougth most voltage values are given against the reversible hydrogen electrode (RHE), this is not a practical reference electrode. The most common reference electrode would then be an Ag/AgCl electrode. However, this might be problematic in solutions with incompatible compounds. In that case, either the electrolyte or the reference electrode must be modified. (Memming, 1998)

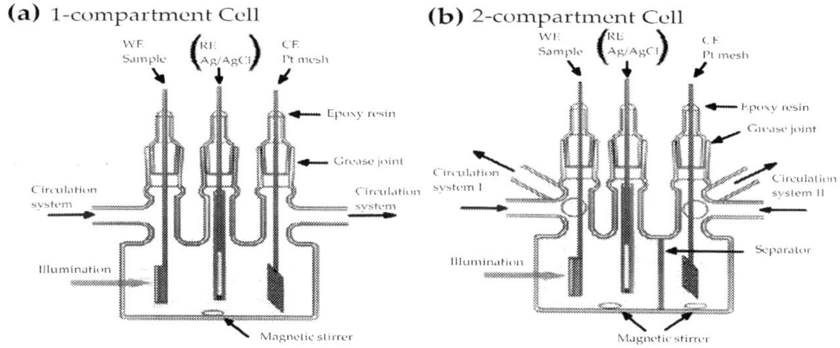

Figure 12. PEC cell with a 3 electrode configuration in a single (a) and double (b) compartment design. The working electrode (WE), counter electrode (CE) and reference electrode (RE). From (Memming, 1998)

Another consideration would be the electrolyte used. First, the material should be stable enough in a certain electrolyte for it to be stable. Once this is satisfied, a neutral electrolyte could be used, although acidic or alkaline media might enhance the kinetics for HER and OER, respectively. In addition, the possible reactions and absorption of each compound at the surface can also change with the species used in the electrolyte. Finally, the concentration of these species can also define the resistivity of the solution, modifying the final performances. An example of the resulting measurements is depicted in Figure 13.

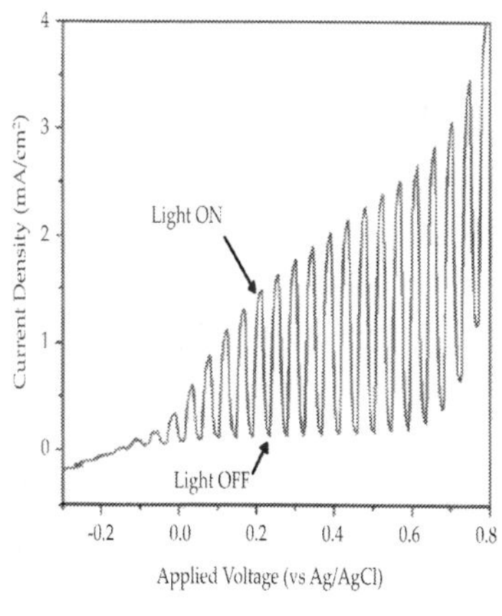

Figure 13. J-V characteristic of a Fe_2o_3-based PEC electrode in 0.5 M KNO_3 at an illumination intensity of 400 mW/cm^2. The light was chopped at a frequency of 0.5 Hz. From (Memming, 1998)

An important aspect when performing these measurements is to determine the potential range for photocurrent generation. That would be affected by the semiconductor type, performance and conductivity. The type of semiconductor would determine the direction of the potential scan. It is also possible to chop the light, to constantly monitor the differences that can appear between dark current and photocurrent.

The scan rate should also be adapted, since high current electrodes would build up a higher double-layer capacitance. Finally, to determine the flat-band potential, both the dark and illuminated current are necessary. Then, the dark current should be subtracted from the illuminated one. If this analysis is repeated at different pH, the Nernstian dependence with pH can be determined. At each pH, the linear portion close to the onset of photocurrent is fitted to a linear line, and the x-intercept is considered as the flat-band potential. (Chen, et al., 2013)

From these measurements, the flat-band potential can be estimated, the dark current could indicate possible degradation, and the operational conditions could be estimated. In order to isolate a photoelectrode performance from the whole systems, a three electrode measurement is normally used. That implies that a reference electrode needs to be used, and the voltages are often reported vs. the reversible hydrogen electrode (RHE), where 0 V represent the reversible potential for HER, and 1.23 V represents the reversible potential for the OER. An example of these measurements is given in Figure 14. In order to determine the operating point, the crossing point between anode and cathode measured under 3 electrode measurements can be used, since both electrodes would need to drive the same current. However, this method often overestimates performance, since the combination of several materials and devices is rarely a perfect process. In addition, it must be accounted that some light will be lost in the process too, lowering the performance. (Chen, et al., 2013)

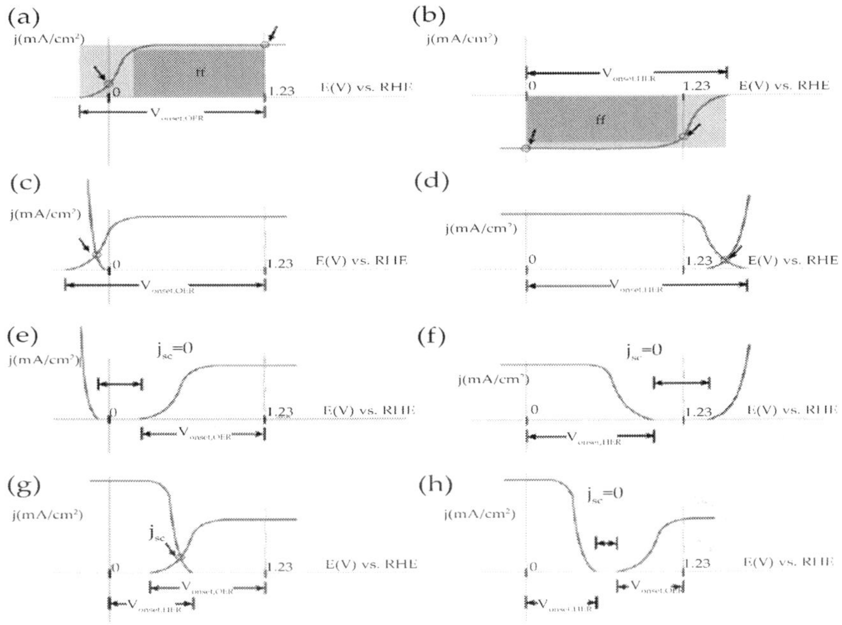

Figure 14. Idealized illuminated cyclic voltammograms for a) photoanode, b) photocathode, c) photoanode with large $V_{onset,\ OER}$ overlaid with dark current from a HER cathode, d) photocathode with a large $V_{onset',\ HER}$ overlaid with dark current from a OER anode, e) photocathode with small $V_{onset,\ OER}$ overlaid with dark current from a HER cathode, f) photocathode with a small $V_{onset',\ HER}$ overlaid with dark current from a OER anode, g) photoanode overlaid with a photocathode, in which the combined V_{onset} is enough to drive the reaction and h) photoanode overlaid with a photocathode, in which the combined V_{onset} is not enough to drive the reaction. From (Chen, et al., 2013)

Moreover, this measurement can give an idea of the limiting mechanism of the overall process. An example is seen in Figure 15, where the potential scan can lead to a current peak at certain potentials. This peak occurs because during the first time interval sufficient redox molecules are available, but in the second scan, the diffusion of molecules to the surface controls the process. (Memming, 1998)

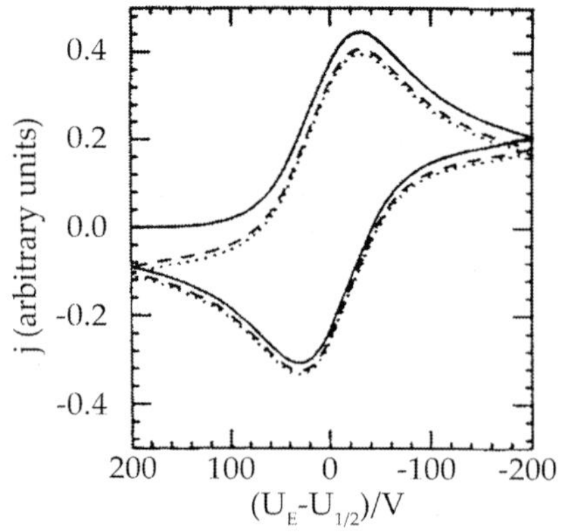

Figure 15. Cyclic voltammetry at various potential scans. From (Memming, 1998)

There are several limits to the accuracy of these experiments. Firstly, the results can slightly vary depending on the reference electrode used. Moreover, the onset potential does not necessarily define the flat-band potential due to the effects of the depletion region at the semiconductor surface, which might change the kinetic overpotential required to drive the reaction. Although the catalyst deposition can minimize these effects, it also introduces one additional variable that can change these properties. Thus, the catalyst would ideally improve the accuracy of the flat-band potential measurement, although it could also result in Fermi level pinning due to changes in the band structure at the surface. Another source of errors might be the difficulty of determining the surface area exposed to the electrolysis. Finally, the effect of corrosion might have an effect in this measurement. Thus, good measurements would need a certain stability during the measurement to be effective. (Chen, et al., 2013)

CHEMICAL CHARACTERIZATION

Ultimately, the objective of a PEC device is to produce the chemical species H_2 and O_2. Thus, it is important to determine its performance not only electrically, but to actually measure the hydrogen production and stability of the samples. The possible methods to do so are summarized in this subsection.

Hydrogen Detection

Direct and quantitative detection of H_2 and O_2 provides evidence of actual water splitting, which would validate the calculations of STH efficiency measurements. From both the electrical and chemical measurements, the Faradaic efficiency can be calculated. Calibrated gas analyzers such as gas chromatography (GC) or mass spectrometer (MS) can be used for this purpose. (Chen, et al., 2013)

The method consists of a reactor, and the evolved gases are quantified by simply measuring the volume of the gases separately in the anode and cathode. However, the identity of those gases must then be confirmed by GC or MS.

Gas Chromatography (GC)

This analysis is normally used for the detection of H_2 and O_2, and other gases produced by other undesired reactions. The equipment consists of a gas injection component and a gas separation component with suitable columns and detectors. The setup is shown in Figure 16. Columns of molecular sieves are used for the separation of the different gases. The particle size of the absorber, and diameter and length of the column would affect the separation process. The columns would need to be regularly be baked when the retention times become too short for adequate separation. (Chen, et al., 2013)

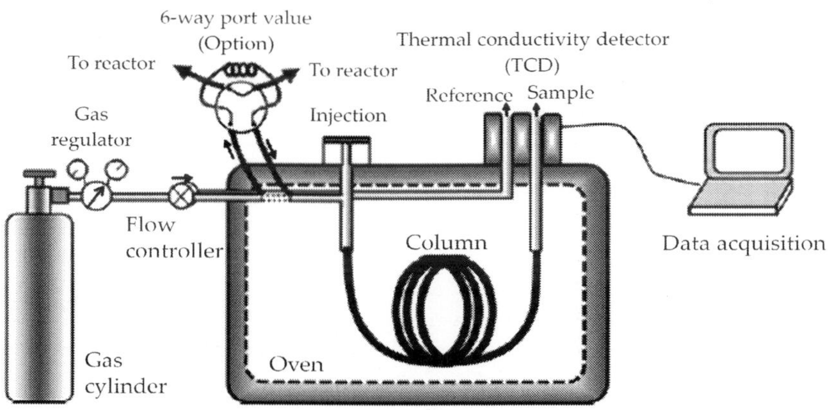

Figure 16. Schematic representation of a basic GC equipped with a thermal conductivity detector. From (Chen, et al., 2013)

Mass spectrometer (MS)

It is a highly sensitive process for qualitative measurements of small amounts of gases. However, it lacks the quantitative accuracy of GC due to the difficulties of a constant gas flow inlet. Thus, careful calibration is needed. (Chen, et al., 2013) However, when used correctly, it can give accurate measurements of the gases produced, being able to determine the faradaic efficiency of the system, as shown in Figure 17. In addition, the stoichiometry of the gases produced can be determined, giving an idea of the side reactions that might be occurring in the system.

Some errors may occur when using these techniques due to the interaction of the product gases with the ambient gases. Thus, a type of reactor that avoid contact between the two is preferred. In order to differentiate as much as possible the two product gases, the collection should ideally occur in each electrode separately, possibly using two different compartments and a separator such as a membrane. In addition, the effects of corrosion of the electrode should always be taken into account. (Chen, et al., 2013)

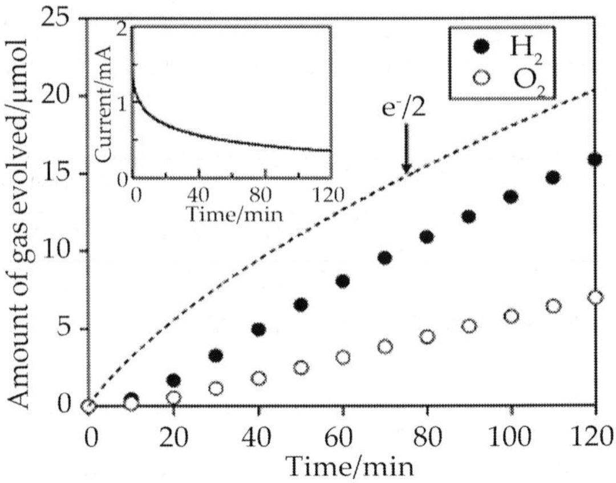

Figure 17. Time evolution of PEC water splitting on a IrO_2/TaON electrode under visible light illumination of a Xe lamp with a Pt electrode in an aqueous 0.1 M Na_2SO_4 electrolyte. Stoichiometric production of H_2:O_2 was observed. From (Chen, et al., 2013)

Stability Tests

As mentioned several times, corrosion of the photoelectrodes is one of the main barriers to realize a fully PEC device. This occurs because photogenerated electrons and holes, with strong oxidizing and reducing potentials, may oxidize or reduce the semiconductor itself.

To evaluate the durability of a semiconductor, the current produced can be measured in time under illuminated conditions, and depending on the reduction of the current, the corrosion can be estimated. These experiments could be performed under no overall bias voltage in a 2-electrode configuration or, more commonly, with a certain bias voltage applied in a 3-electrode configuration. For a more thorough understanding of the nature of the corrosion, additional characterization such as optical or electron microscopy, X-ray photoelectron spectroscopy (XPS), X-ray diffraction (XRD) or Raman spectroscopy could be used, measuring before and after corrosion. (Chen, et al., 2013)

Some aspects that must be taken into account are the stability of the lamp over the time period of the test, the possible heating occurring due to long light exposure, and possible contaminations during the experiments. To avoid these problems as much as possible, the experiment should be closely monitored. Finally, these test are rather time consuming, since common standards require testing for 1000 h. Thus, future testing may require more sophisticated and accelerated testing. (Chen, et al., 2013)

SOLAR SIMULATOR CONSIDERATIONS

As shown in several of the measurement and characterization techniques, the light source used plays an important role on the obtained results. Thus, it is crucial to better understand the challenges and inaccuracies that can stem from incorrect light sources.

First, it is important to understand what it is considered the reference spectrum for solar light. As previously explained, the standard spectrum, AM1.5, is considered to have an intensity of 1000 W/m^2 and a wavelength distribution as shown in Figure 18. This is chosen because it represents an average of the usual values of the solar irradiance. In addition, the combination of direct and diffuse irradiance must also be taken into account. The global irradiance is assumed to have 90 % direct and 10 % diffuse irradiance. (Chen, et al., 2013)

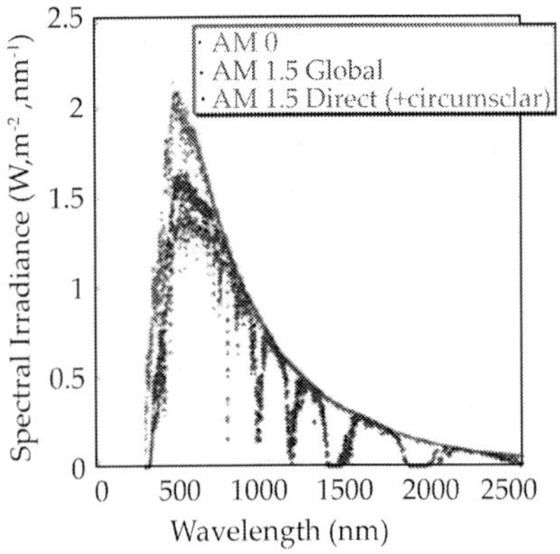

Figure 18. AM 0, AM 1.5 global and AM 1.5 direct spectra plotted using atmospheric data. From (Chen, et al., 2013)

Different lamps together with filters can be used to achieve this spectrum. Figure 19 shows the spectrum of different lamps. Filament lamps are often preferred due to their lower cost and more stable output intensity. However, the infrared (IR) part of the spectrum is too high, and thus some filter must be used to eliminate that part of the spectrum, and cooling water should also be used. (Chen, et al., 2013)

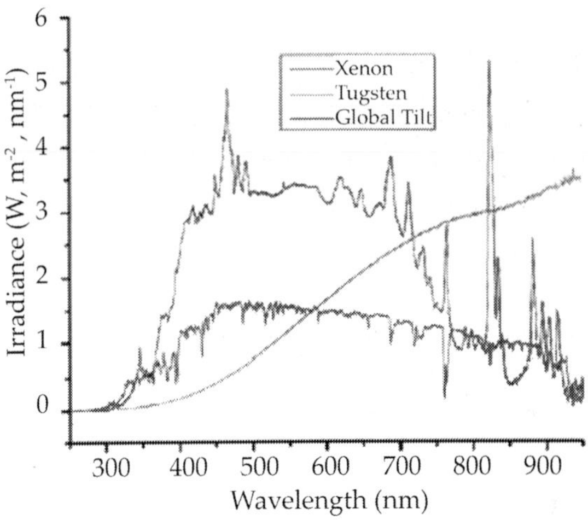

Figure 19. Spectral output of a 1 kW Xe-arc lamp and a 250 W tungsten-halogen lamp as measured using a Si spectroradiometer, and calibrated against the AM 1.5 G spectrum. From (Chen, et al., 2013)

Commercial simulators usually use arc sources, filtered to mimic the AM1.5 G spectrum, although they do not replicate it exactly. These simulators are usually calibrated to reach the short-circuit current equivalent at that under real AM1.5 G spectrum, as illustrated in Figure 20. That is usually done using reference cells that have been previously calibrated with a known spectrum. These cells can also be used to assure the stability of the solar simulator over time. To approximate this spectrum more closely, some solar simulators include two lamps, one of xenon and one of tungsten, since have a better spectrum at low wavelengths (Doscher, et al., 2016), while other are better at high wavelengths. That also assures that the simulator is valid at all bandgap semiconductors. (Chen, et al., 2013)

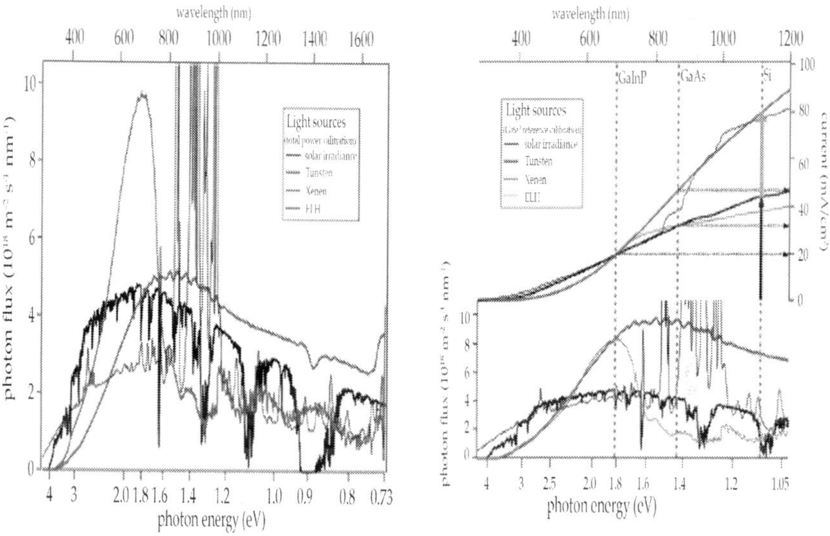

Figure 20. Current limitation vs. absorber bandgap for AM 1.5 G and various laboratory white-light sources (top), based on a GaInP reference cell, and the photon flux at each wavelength for each of these sources. From (Doscher, et al., 2016)

Spectral mismatches and incorrect calibration has resulted in incorrect performance measurements in numerous reported PEC devices, such as the GaInP/GaAs with a reported performance of 18 %. (Licht, et al., 2001) It was later reported that critical evaluation of the illumination light source used revealed a significant systematic error that highly overestimated the STH efficiency of that device. (Doscher, et al., 2016) Therefore, these studies show that calibration of the solar simulators is a crucial step when characterizing PEC devices. In addition, these results should be complemented by other characterization techniques such as IPCE, and reflectance and transmittance measurements, to ensure that the estimated STH efficiency and other performance indicators are correct. Finally, it is important to consider the exact illuminated area and the possible artificial reflections within the system to avoid overestimation of the current density produced within the PEC device.

SUMMARY

When reporting and comparing PEC performances, it is of vital importance to know how these devices were measured, and identify possible areas of inaccuracies. The characterization has been divided in three main categories: optical, electrical and chemical characterization. Regarding the optical characterization, it is mainly used to determine the absorption, the performance per wavelength and the structural characteristics via SEM.

The electrical properties are mainly characterized by the onset potential and the J-V response. In the first case, the potential produced in dark and illuminated conditions are compared. In the case of the JV characteristic, it consist on a voltage sweep while measuring the current density under solar light illumination. There are several things to consider in this measurement. First, a reference electrode could be included if the measurement is targeted at one of the two electrode. Otherwise, a two electrode configuration should be used. Another important parameter to take into account is the possible degradation and side reactions. To help determining this, the chemical measurements can be done, measuring H_2 and O_2 and calculating the stochiometry of these reactions, and stability measurements can be done to look at the current production over time.

The final parameter that can affect the measurements is the used light. There are several types of lamps and filters that would need to be used to achieve a spectrum as close as possible to the AM1.5 G. In addition, calibration would be needed by using a Si diode. This is very important to be able to compare different measurement along different labs.

CHAPTER 7

ECONOMICS AND BUSINESS PERSPECTIVES

Intensive research in the last years was focused in using more earth-abundant materials while still achieving high efficiencies, so that these devices can be applied in a bigger scale with economic and financial advantages. (Turan, et al., 2016) This is necessary due to the intermittency of the solar resources, which would otherwise have to be covered with other technologies such as batteries or flywheels. But these technologies do not have the energy density that chemical fuels, and thus the latter seem to have the most potential for long-term energy storage. However, in order to determine the economic feasibility, there are some practical issues within the design of PEC devices that need to be considered, and would be addressed in this chapter. In general, the target cost of hydrogen production of approximately \$1.60-10.40 per kg H_2 at commercial scale, being then cost competitive with fossil fuels. (Pinaud, et al., 2013)

MATERIALS AND CELL CONFIGURATIONS

The first factor that needs to be taken into account, and was already previously discussed, is the materials that would be used. That would depend on the bandgap, electrical properties and band alignment to which efficiencies would be able to reach.

In addition, it was discussed that in order to make an autonomous system, several junctions, and thus several materials would be needed. Depending on the materials used, the costs and the efficiencies of these system would be determined. The maximum STH efficiencies for a dual absorber system are displayed in **Figure** 1 and **Figure** 2 in the case of stacked together and placed side by side, respectively. (Pinaud, et al., 2013) It seems from this analysis that a stacked configuration would result in much higher efficiencies, since the area needed for such a device would be much lower than if the y are placed side by side.

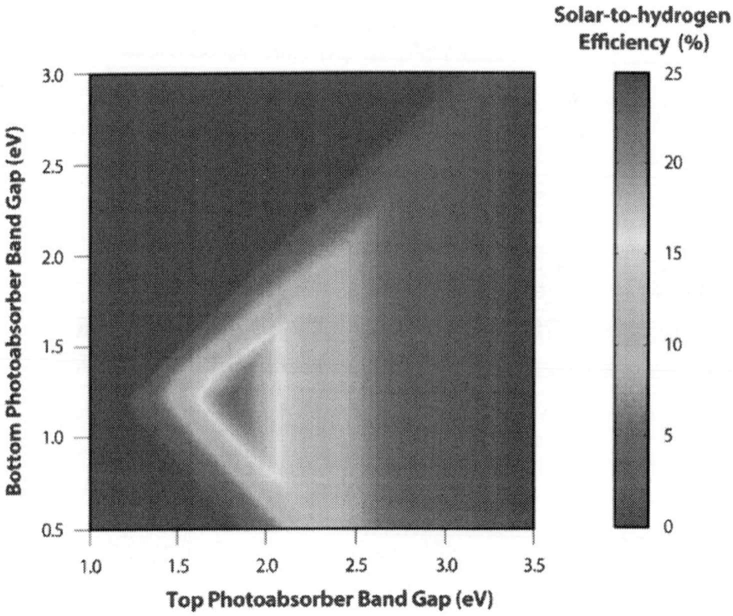

Figure 1. Maximum theoretical solar-to-hydrogen efficiency for a dual stacked absorber configuration plotted as a function of the top and bottom photoabsorber band gaps. The top photoabsorber is assumed to be placed above the bottom photoabsorber, thus only photons with energy less than the band gap of the former are transmitted to the latter. From (Pinaud, et al., 2013)

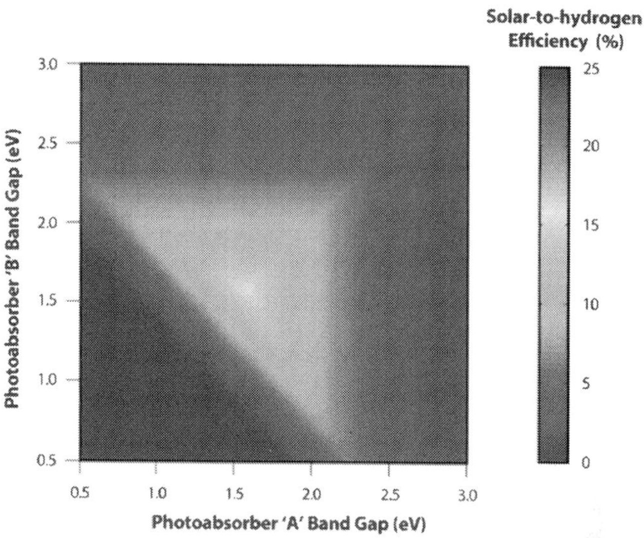

Figure 2: Maximum theoretical solar-to-hydrogen efficiency for a dual side-by-side absorber configuration plotted as a function of the two photoabsorber (denoted 'A' and 'B') band gaps. The two electrodes are assumed to be placed next to each other and can each access the full solar spectrum, from (Pinaud, et al., 2013)

The maximum efficiency for the stacked case is 22.8 %, with absorbers with 1.23 and 1.84 eV bandgap absorbers, as opposed to the side-by-side configuration, in which a 15.5 % with two absorbers of 1.59 eV. These values are lower than the thermodynamic limit of 41 %, which is merely based on the bandgap absorption characteristic of AM 1.5 G. These simulations, on the other hand, have also realistic assumptions regarding the catalytic activity and electrical performance, rather than just the pure optical absorption. Nevertheless, even though these STH efficiencies are far from the thermodynamic limit, they are still very promising results, since they are efficiencies higher than 10 %. (Pinaud, et al., 2013)

Thus, the main determinant parameters to choose the materials to use in a commercially viable device are the following:

- High photon absorption
- High photocurrent production
- Bandgaps between 1.5 and 2.2 eV to match the needed potential for the reaction.

- Low cost and earth-abundant material
- Durable in an aqueous solution
- Resistance to high solar irradiance and high temperatures
- Able to support high current flows
- Avoid poisonous or toxic compounds and contaminant materials
- The electrolyte must also be non-toxic and non-corrosive to avoid any hazards.

Considering the configuration and cell reactor design, there are several possibilities with different degree of complexity. The choice would depend on the budget and on the device architecture. Different possibilities of the cell configuration, and the translation from lab scale devices to larges scales is depicted in **Figure** 3. Once the desired cell configurations chosen, it can be replicated to form a reactor, which would be explained in the next section.

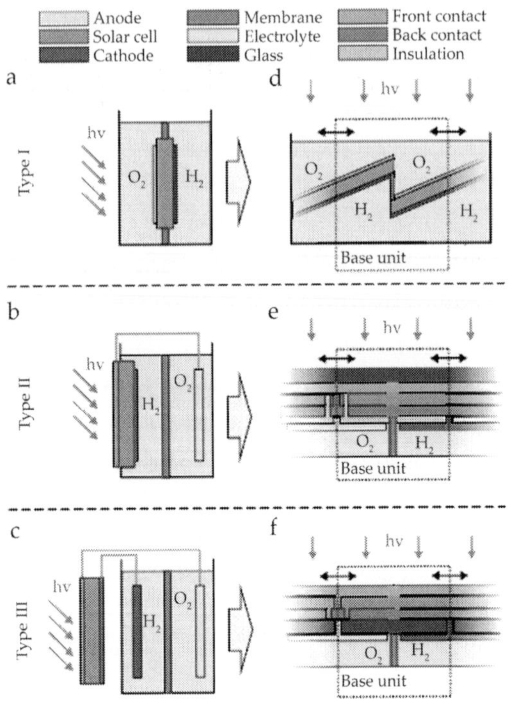

Figure 3: Classification of different photovoltaic water splitting design. The lab devices are shown in the left (a-c), while their translation into the respective cell

reactor configurations are in the right (d-f). Type I is the PV device completely immersed, Type II includes only one side of the semiconductor in contact with the solution, and Type III completely separates the semiconductor from the electrolyte, in this case in a wireless configuration. From (Turan, et al., 2016)

REACTOR DESIGN

If PEC are to be applied to a commercial scale, reactor design would come into account. Several designs have been proposed pictured in **Figure** 4. These are of course not all the possible configurations, but only a sample range, which would allow for a technical and economic evaluation. Each system includes all the elements to go the whole process from absorbing the solar light to generate the voltage and the HER and OER, so that the PEC system becomes completely independent. (Pinaud, et al., 2013)

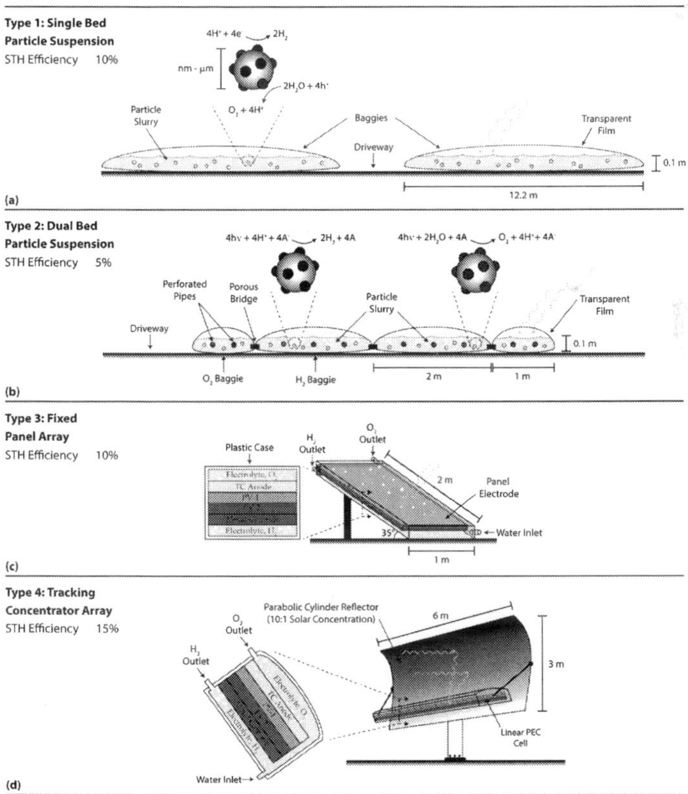

Figure 4: Schematic of the four reactor types including (a) Type 1 reactor cross-section showing the particle slurry contained within baggies separated by an

access driveway, (b) Type 2 reactor cross-section showing the particle slurries contained within baggie assemblies consisting of an alternating arrangement of a full size and half-size baggie each for O_2 and H_2 evolution, (c) Type 3 reactor design showing the encased composite panel oriented toward the sun with buoyant separation of gases, and (d) Type 4 reactor design with an offset parabolic cylinder receiver concentrating light on a linear PEC cell. Drawing not to scale. From (Pinaud, et al., 2013)

There are two main types of reactors, the enclosed reactor beds with suspended photoactive particles and the multilayer arrays immersed in aqueous electrolytes, which should be oriented towards the sun.

Each reactor will be now explained in more detail:

- *Single bed particle suspension reactor*: This is considered the simplest reactor of all of the ones shown here. It consists of a low lying transparent bag containing a slurry of photoactive particles in a conductive electrolyte. The bag is usually made out of high density polyethylene (HDPE) due to its transparency, low hydrogen permeability and resistance to degradation in an alkaline electrolyte, and low cost. The size of the bag, in particular the length and width, can be optimized to make the area as large as possible while still allowing the bag to be made by single pieces of plastic. When gas is generated by the evolution of both half reactions and drawn out for further use, the upper portions of the bag reactor will rise and fall. Thus, the gas would be accumulated on the top of the reactor during the day, when the light is shining, and removed at regular intervals. Moreover, since the bag is completely sealed, no electrolyte is lost. However, the temperature should be maintained at slightly higher temperature than ambient, so that the thermodynamic voltage is slightly lowered, but at the same time the electrolyte is maintained in liquid form. Another important material to optimize is the size and shape of the particles. (Pinaud, et al., 2013)
- *Dual bed particle suspension reactor*: This is similar to the previous reactor. The primary difference is the use of separate beds for the two half reactions, HER and OER. This has several advantages: intrinsic separation of the two gases, lowering the costs of separation; and more flexibility in the choice of

semiconductor since the beds for the two half reactions are separated. However, this system needs a redox mediator to interchange ions between the two compartments, so that the reaction can be completed. This can be any species with a fast and reversible reaction and large diffusivity, such as iodine, bromine or iron complexes. The bridges for these ions consist of membranes or gas/liquid interfaces, allowing the ions to exchange but not the produced gases. The dimensions of these bags must be optimized to minimize the distance between the two complementary reactions. The efficiency that can be achieved by this system is lower than the first system, since the semiconductors are placed side by side. However, the practicalities of this system can achieve a practical higher efficiency and lower costs of separation. (Pinaud, et al., 2013)

- *Fixed panel array reactor*: This system has been among the most researched topics to date regarding PEC reactors. These products also resemble the commercial photovoltaic modules, which might help with the commercialization of this product. In this system, the multiple photoactive layers are sandwiched in between the two electrodes, and the whole structure is encapsulated in a transparent case that contains the electrolyte reservoir. The bottom of the stack would then be composed of a metal cathode where the electrons are collected to drive the HER reaction. Gas separation is achieved by physically separating the two electrodes, collecting the gases at the top of the panel. The panels would be tilted towards the sun to maximize the overall yearly energy collection. (Pinaud, et al., 2013)
- *Tracking concentrator array reactor*: This reactor is similar to the previous one. However, a concentrator system, as well as a tracking system have been integrated. These devices are able to maximize the direct radiation that falls into the PEC module, allowing for higher efficiencies and higher overall performances with less material costs. Within concentrators systems, parabolic cylinder arrays are often used, placing the planar array on the focal point to maximize the light falling into the module. The advantages of these concentrators over

other systems are their low cost, relatively low weight, and the convenience to place the gas and electrolyte pipes on the base of the reactor. Moreover, the ideal concentration ratio of PEC systems has been determined to be between 1:10 and 1:100, due to limitations in the current densities that catalysts are able to handle. On the other hand, the most adequate tracking system in this case is judged to be a double axis tracker, allowing for both changes in the tilt and azimuth of the panel to better track the movements of the sun during the day. It is important to know that concentrators are often relatively costly, and thus this approach is only reasonable with high-cost semiconductors such as the ones based on III-V technologies. (Pinaud, et al., 2013)

However, these reactor designs also suffer from limitations. The first two reactors have higher uncertainties regarding the photoactive material fabrication and performance compared to the planar cases, and the demonstrated efficiencies are much lower. In addition, there are uncertainties in these cases with regard of the areas for absorption and reactive surfaces. These particles are also covering a certain area, possibly creating mutual shading between them, resulting in a lower effective incident light intensity for the bottom particles. Finally, the integrity of the bags might be compromised, creating hazardous situations. In particular, the first type of reactor can create dangerous situations if the non-trivial separation of the produced hydrogen and oxygen is not totally successful. Thus, these technologies are considered to need further investigation and developments to achieve the desired performances. (Pinaud, et al., 2013)

Regarding the last two reactor types, they are much more complex than the first two. In general, they would create higher current densities than the previous reactors. Thus, also higher requirements are placed in the catalysis since it has to redirect a larger amount of charge carriers to drive the reactions. This can also be done by increasing the catalyst area while maintaining the absorber area. However, decoupling these two areas would increase the complexity and costs of the system. Moreover, this decoupling might create a voltage drop due to losses in additional cables and other elements. (Pinaud, et al., 2013)

In addition, an economic analysis of these reactors has been performed by (van de Krol & Gratzel, 2012), as shown in **Figure** 5. The conclusions of this analysis reaches similar conclusions than the previous report, being the flat planar the most costly system, with maintenance and lifetime being the most important costs that need to be considered, while the materials play a very important role into other reactor types, especially the bag type reactor. (van de Krol & Gratzel, 2012)

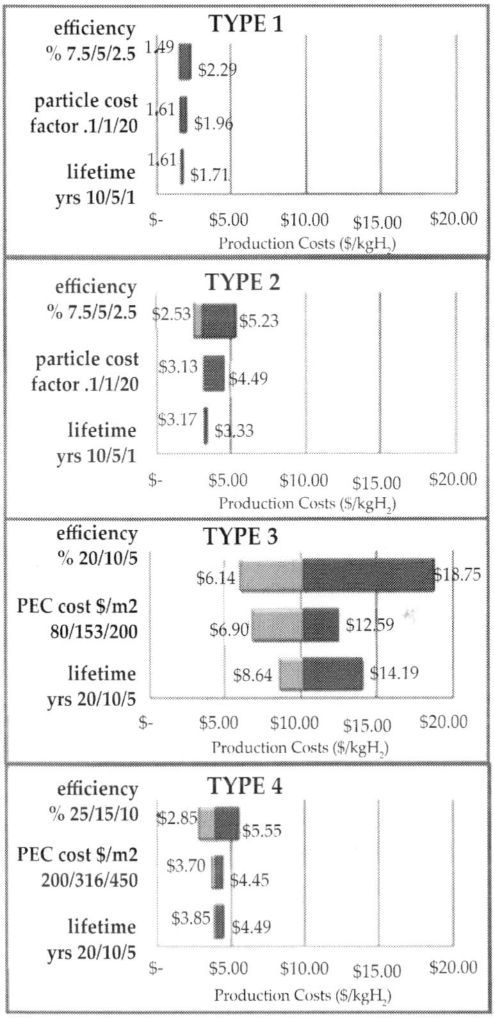

Figure 5: Hydrogen cost sensitivity analysis for the four reactor types previously described. projected hydrogen costs as a function of efficiency, lifetime and material costs are shown. From (van de Krol & Gratzel, 2012)

Nevertheless, reactor design is not the only factor influencing the final cost of hydrogen production. There are many elements such as pumps, pipes and separation equipment that also have to be taken into account. Each reactor type, according to their design and specific needs, will have a different set of elements surrounding its use. Thus, due to the importance of these elements, the following chapter discusses the possibilities and economic consequences of overall plant design.

PLANT DESIGN AND OPERATION

The design of a full plant for PEC operations would be based on the dimensions, operating conditions and performance of the chosen reactor, as described in the previous section. In addition to the reactor, the plant would also include all the gas separation units, purification devices and compression equipment to deliver the final product, often compressed hydrogen gas. In addition, also the layout and spacing must be accounted for, as well as supporting piping, pumps and wiring. (Pinaud, et al., 2013)

It must be taken into account that the performance of the final plant would also depend on the solar insolation of the chosen location, and on the daily and seasonal fluctuations that might appear. Ideally, it would be located in a sunny location, facing the equator at the right angle, and with no possible shadow from nearby buildings, trees or mountains. Once all these parameters have been optimized, the final insolation received by the reactor would then depend on the type of reactor used at the chosen location. **Figure** 6 shows the average monthly insolation at a certain location depending on the reactor type. Thus, this shows that the tracking concentrator system can achieve the higher insolation, since it is able to track the sun, while the bag-type reactors achieve the lowest insolation due to the partial shading between particles and the light absorption in the plastic of the bag. Interestingly, the fixed array panel can be con**Figure**d in a way that achieves a similar performance as the tracking system in winter, and although the production is relatively low in summer, it is able to maintain a constant production of hydrogen throughout the whole year. Ultimately, the average insolation of each of these systems would determine the area needed to cover certain flow of activity and production targets. Thus, the less efficient systems, which are also the cheapest ones, would in turn need the biggest surface areas, creating a trade-off between technological cost and land use. (Pinaud, et al., 2013)

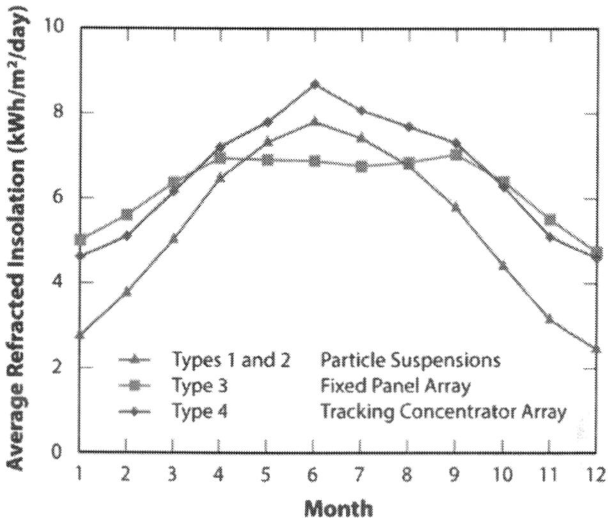

Figure 6. Average monthly refracted insolation for each reactor type. Refracted insolation refers to the light intensity incident upon the photoabsorbers, as opposed to the incident intensity (not shown), which is incident upon the HDPE or PMMA reactor covering. From (Pinaud, et al., 2013)

As previously mentioned, once the gas is produced in the reactors, it needs to be separated, purified and compressed. Most of these elements are independent of the reactor used. However, in some instances the type of reactor can affect the needs in post processing. For example, only the single bed reactor needs thorough purification of the products, since it is the only reactor in which all the products are simultaneously produced in the same chamber. Some common systems for hydrogen separation are pressure swing absorption (PSA), temperature swing absorption (TSA), palladium and nano-porous membranes, and electrochemical pumps. Moreover, piping and wiring are needed. It must be noticed that hydrogen can make many materials commonly used for piping brittle, such as stainless steel. Thus, more advanced materials like polyvinyl chloride might be needed. This would also be affected by the operation temperature and pressure of the different systems, and thus operation conditions must be carefully chosen. After purification, the hydrogen would first need to be cooled to about 40°C, and then it can be compressed using a system such as a two-stage piston compressor. Maintaining control of the temperature during these steps is very important, since hydrogen is a highly volatile and hazardous gas, which could create a dangerous situation when overheated. Finally, a control system is also needed for monitoring purposes, alarming for hazardous conditions and controlling all plant equipment for smooth operation of the PEC plant. To minimize control costs, it is recommended that the complexity and sophistication of the plant designed allows complete functionality and safe operation while still reaching a level of simplicity that can allow for more straight forward control systems, which would in turn lower the costs of operation and monitoring. (Pinaud, et al., 2013)

Finally, it must be considered that, depending on the type of reactor, the energy and water needs would be different, creating again a trade-off between the two cases. The estimated uses for the energy and water needed, as well as the equipment where it is used, are summarized in Table 15. (Pinaud, et al., 2013) It can be seen that Type I needs a higher amount of energy due to the separation process. On the other hand, the Type IV uses hardly any energy in comparison to the others. Regarding the water use, Type I uses slightly more energy than the others, both due to the electrolysis process and vapor loss. Thus, in this sense, Type IV would be the preferred configuration, as it has the least energy and water consumption.

Table 1: Utility usage for unit operation and feedstock for producing each of the four reactor types. From (Pinaud, et al., 2013)

	Type I	Type II	Type III	Type IV
Power consumption				
Compressor (kW)	128.6	76.3	76.3	-
Pumps (kW)	1.1	0.9	0.7	0.22
PSA (kW)	1.0	-	-	-
Control system (kW)	6.5	6.5	6.5	6.5
Average power consumption (kW)	137.2	83.7	83.5	6.7
Water consumption				
Electrolysis (kg per day)	9928.8	8935.9	8935.9	8935.9
Vapor loss (kg per day)	53.1	31.9	31.9	31.9
Total usage (kg per day)	9981.9	8967.8	8967.8	8967.8

It is important to also consider the problems that might arise from the scaling up of the PEC device that have been discussed and researched throughout this book. Research is usually carried out in small samples in the order of millimeters to centimeters. However, in larger scale, the sources of resistances and limitations might change since the travelling distances of the compounds would change. In addition, the amount of electrolyte might change, also varying its electrical resistance and light absorption. Moreover, the current flows and collection mechanisms have to be taken into account, which has not been researched in depth yet. (van de Krol & Gratzel, 2012) Thus, these aspects must be explored when a plant is constructed.

ECONOMIC AND FINANCIAL CONSIDERATIONS

To evaluate the economic feasibility of building such a plant, a detailed analysis was conducted by (Pinaud, et al., 2013). 2 summarizes the most important assumptions taken by the authors of that report in order to determine the estimates presented here. When doing these calculations and assumptions, it is important to keep in mind that implementation can be more costly at the moment, since there is no infrastructure at the moment such as a hydrogen distribution network, and thus the cost of delivery outside the plant would be relatively high. However, these costs have not taken into account in this analysis, and the presence of a hydrogen distribution network has been assumed instead, since this book focuses on the sustainable production of hydrogen rather than the total implications of implementing a hydrogen economy. (Pinaud, et al., 2013)

Table 2. Main assumed parameters to calculate the costs of the overall plant design. From (Pinaud, et al., 2013)

Operation period and facility life	20 years
Construction period	1 year
CO_2 capture credit, taxes or O_2 credit	None
Depreciation type	Modified accelerated cost recovery system, 20 years
Inflation rate	1.9 %
Land cost	$ 500 per acre

Hydrogen pressure at central gate	300 psig
Hydrogen purity	98 % minimum, CO < 10 ppm, sulfur < 10 ppm
Production facility maintenance and repair	0.5 % of direct capital cost per year
Burdened labor rate for staff	$ 50 per hour
Operating capacity factor	90 %
Site preparation	1 % direct costs minus unique excavation costs
Engineering and design	7 % of direct costs
Process contingency	20 %

As could be expected, the costs estimated within this study vary depending on the type of reactor considered. In addition, the location is also important, since not only determines the solar insolation received by the system, but also the cost of the land, electrical energy and employees. The plants considered in this study produced a volume of 41 600 kg of hydrogen per year, with the main costs of the bag-based reactors located on the land use, while for the planar configurations the systems needed surrounding the reactor are the more costly element. The analysis concludes that, based on these systems, a cost of $ 304 per kg of hydrogen is needed, where about two thirds of it is due to the costs of materials, and the rest is directed to the coating process and, especially, for markup, R&D and profit. (Pinaud, et al., 2013) This is obviously too expensive to be practical, and thus new technologies and configurations are needed to reach the cost-competitive levels desired. A factor that must be considered when comparing these systems and their costs is that the bag- based systems have a relatively low lifetime, with an estimation of one change per year for both the bag and the particles, while the planar devices are expected to last for 20 years.

Another factor to take into account when designing these systems is the scale at which they are implemented. An increase in scale leads to a decrease in labor per kilogram, also reducing the costs. Thus, having a centralized facility for solar hydrogen production can lead to cost reduction. The levelized cost of hydrogen for each reactor type previously described are \$1.60 per kg H_2 for type 1, \$3.20 per kg H_2 for type 2, \$10.40 per kg H_2 for type 3, and \$4.10 per kg H_2 for type 4. The goal to achieve is a cost of approximately \$2.00-\$4.00 per kg H_2, which according to this study seems to be an achievable goal. In addition, educated comparisons can be made between the different systems, concluding the single reactor bed to be the most economically efficient, although its performance in terms of hydrogen conversion would be the lowest. For a more detailed analysis, a distributed analysis of the costs is depicted in **Figure** 7, whereas expected the capital costs and fixed operation and maintenance are the most important. (Pinaud, et al., 2013) It is important to notice the high costs of Type III, due to the high capital costs. In addition, despite the additional separation processes needed, Type I seems the most economically feasible.

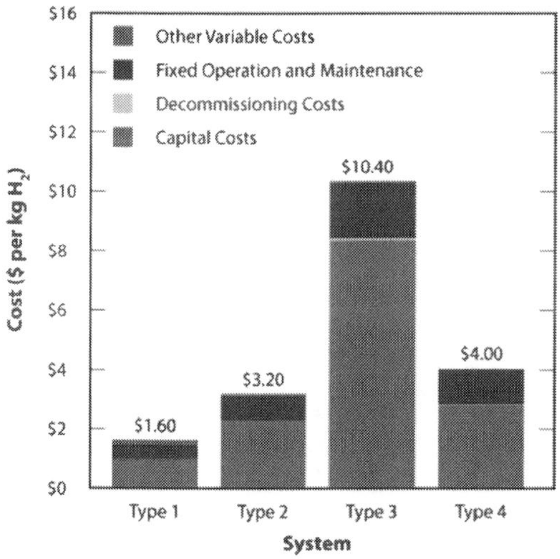

Figure 7. Distribution of cost contributions to the levelized price of hydrogen. From (Pinaud, et al., 2013)

Furthermore, to determine the elements to focus on in order to bring these costs even further down, a sensitivity analysis was conducted for all reactor types and configurations. The results are summarized in **Figure** 8, along with the main parameters studied. In all cases, the capital costs seem to bring the biggest costs to the system, this is a positive conclusion since new developments and inventions could easily bring this value down. It is also important to notice that the design uncertainties at this stage could also lead to additional costs that are not counted for here. It is also interesting to notice the big portion of the investment going to purification and compression systems in the case of a single bed bag reactor, which determines the relatively high costs. (Pinaud, et al., 2013)

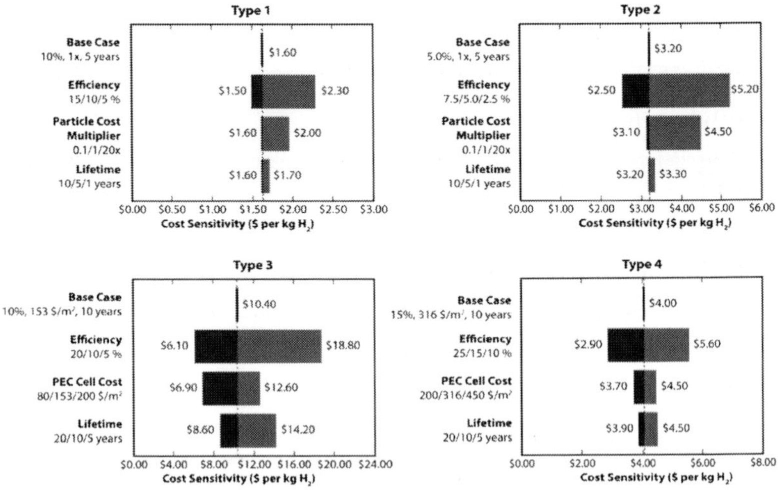

Figure 8: Effect of efficiency, particle or panel cost, and component lifetime on the cost of hydrogen from each reactor design. Each calculation represents the variation of a single parameter from the base case to a higher and lower value as indicated on the left axis. From (Pinaud, et al., 2013)

There are other sources of possible additional costs, such as the increased safety measures and monitoring systems needed when hydrogen is compressed, since this is a highly reactive and hazardous material that would need special measures. On the other hand, the concentrating systems can also significantly raise the costs of the system to the point where, despite the additional efficiency gain, the benefits are not enough to compensate for the loss of capital investment. Regarding the fixed planar array, being the most expensive of all according to the levelized cost of hydrogen, the main costs come from the materials used, construction and installation, over 80 %. In addition, the encapsulation also plays an important role since it needs to be tight enough to avoid any electrolyte leakage. Thus, the costs could be dramatically reduced if more earth-abundant materials could be used as the bas for these devices. (Pinaud, et al., 2013)

Others authors have arrived to an estimated cost of $20/kg without including the land cost, maintenance and installation. The values found with this estimation is clearly too high, and thus further research and improvement of the PEC devices is needed. Otherwise, other estimates suggest that the combination of PV and electrolyzers could be more economical, with levelized costs of hydrogen between \$2-8/kg, considering only the investment costs. (van de Krol & Gratzel, 2012)

Another important factor to consider in the economics of these plants is the durability of the materials, which can be an issue as previously seen. That is especially taken into account in pay-back calculations, since the consequences of instabilities will be seen further in time. The targets of the industry is to have a 10 year payback and 20 years of device lifetime. However, tests of more than 1000 h are rare in literature, and thus more information is needed in this aspect. (van de Krol & Gratzel, 2012)

When considering the use of hydrogen as an energy storage method, it is also important to consider the methods that could be used to store this gas. This can be done in gas cylinders, usually pressurized, or even in liquid form at very low temperatures and high pressures. However, the container is usually very heavy and bulky. More modern and light containers can be fabricated on fivers or special resins. Other ways to store the hydrogen is to further convert it in other compounds such as an ammonia, methane or other compounds. Moreover, if these systems are to be used, infrastructure to be able to transport it easily would be needed, thus the possible brittleness that hydrogen causes must be taken into account. That could increase the costs to achieve a convenient distribution system. The alternative would be to further produce a fuel such as methane, for which a distribution infrastructure is already in place. (van de Krol & Gratzel, 2012)

SUMMARY

Once the main scientific aspects of PEC devices have been reviewed, the economic aspects come into play. This section deals with the scalability of these technologies from lab cells to full operational plants. In this process, there are several approaches that could be taken into account. First, regarding the cell design, there are several possibilities presented showing different surface exposures to the electrolyte. In addition, the translation of these lab approaches to an industrial cell level was discussed.

Then, the reactor design was considered. Four main types of reactors were identified: single particle bed, double bed, planar type and planar concentrated type. When choosing the appropriate reactor, it was outlined that the bed reactors tend to be cheaper, since very simple polymer bags and the particle bed are the only needed components. However, in the case of single bed reactors, additional costs would appear at the plant design level due to the production of the different gases in one compartment, which would require further use of separation technologies, increasing the costs. Planar reactor configuration have higher costs per kg of H_2 produced, especially in the case of the simple non-concentrated one, due to the high cost of the semiconductor and relatively low efficiencies that would be obtained.

Overall, considering the complete plant design and subsequent maintenance costs, it appeared that the single bed reactor yielded the best results in terms of costs, giving a price of $ 1.6 per kg of hydrogen. On the other hand, simple planar configurations reach the highest costs due to relatively low efficiencies and, at the same time, high semiconductor use, coming to a price of $ 10.60 per kg of hydrogen.

CHAPTER 8

CONCLUSIONS AND OUTLOOK

This book has shown the importance of considering all aspects of a technology. Regarding the solar technologies, the storage options must also be weighted. In order to solve the problem related long-term storage that solar energy would face when applied at large scale, energy storage in hydrogen or other chemical fuels can be considered. A valid and direct option to produce such fuels by using solar energy is to apply photoelectrical (PEC) water splitting devices. This book has reviewed the main materials used for such an application, mainly Fe_2O_3, TiO_2, o more modern materials such as WO_3 or $BiVO_4$.

These materials have several characteristics that must be fulfilled:

- Low cost and high availability of the semiconductor materials that would be used as photoelectrode. In addition, they must also have the appropriate bandgap to operate under solar radiation, the correct band alignment to drive the electrochemical reactions, and it must be durable in water, preferably in slightly acidic or basic conditions.

- The interface between the semiconductor and the electrolyte must be precisely studied in order to minimize the overpotentials needed to transfer the charge carriers form one to the other. This would include the optimization of the band alignment, but also the inclusion of an active and effective catalyst, and it may include some sort of passivating layer to avoid the corrosion of the photoelectrode. Innovative ideas can also be included in this surface to further aid the reaction, such as nanostructures to extend the surface area, or the use of quantum dots to manipulate the quantum phenomena happening inside the semiconductor lattice.
- The inclusion of multijunctions can facilitate the charge carrier separation and collection for several reasons. First, the band alignment of the photoelectrode would now concern only one half-reaction, opening room for many materials as candidates. In addition, since these junctions would be placed in series, the voltages of each cell would be added to finally reach the required voltages to drive the water splitting reactions. These multijunctions can be of many natures, such as connecting two PEC photoelectrodes together, or connecting them to an independent solar cell, or a multijunction direct PV device could be considered. However, the absorber materials used and tunneling recombination junctions would have to be further studied and specifically optimized for this specific application.
- New synthesis and fabrication techniques such as ALD can lead to a more homogeneous layer which would result in higher efficiencies and better protection of the photoelectrodes against corrosion. In addition, other modern techniques can lead to better scalable technologies, which would make the application of these methods at large scale more effective and efficient.

Efforts are being made to address these challenges and technical barriers for PEC implementation. The focus of this research is to improve the materials and fabrication techniques used, with special focus on understanding and improving the semiconductor/electrolyte junction.

Nanostructured materials are especially popular on that respect, yielding positive results so far. New and old materials such as tungsten oxides, iron oxides, vanadates and titanium dioxides have been significantly improved by introducing such nanostructures, as well as by doping techniques.

When including the state-of-the-art semiconductors in a large scale reactor, some economic analysis can be drawn. First, it is important to note that there are several types of reactor considerations that can be envisioned, the most important being either bag based with a bed of semiconductor particles, or planar based, with the array in a configuration very similar to a PV system, and which might or might not include sun tracking systems and light concentrator systems. According to the reviewed analysis, it seems that the bed reactors are the most cost-effective due to the low costs of the reactor materials themselves. However, the highest output scan be obtained with the planar systems. In addition, when applying PEC technologies at large scale, not only the reactor needs to be taken into account, but also the separation and purifications systems, as well as the piping, temperature control, etc. In that sense, planar systems have the advantage of separate electrode, resulting in already separated gases and thus eliminating several separation gases steps. On the other hands, these systems are more complex, and especially in the case of concentrating light and sun tracking, more monitoring is needed, increasing the costs of the overall system. Thus, the scalability of each of these reactors and devices must be further investigated to better determine which would be the optimal configuration.

REFERENCES

1. Abdi, F. F. et al., 2013. Efficient solar water splitting by enhanced charge separation in bismuth vanadate-silicon tandem photoelectrode. *Nature communications,* Volume 4, p. 2195.
2. Abdi, F. F. & Krol, R. v. d., 2012. Nature and Light Dependence of Bulk Recombination in Co-Pi catalyzed BiVO4 photoanodes. *Journal of Physical Chemistry C,* 116(17), pp. 9398-9404.
3. Abdullah-Al-Mammun, M., 2010. *Development, Evaluation and Application of Metal Nanoparticles, Nanophotocatalyst and Nanocomposites.* s.l.:Khulna University of Engineering and Technology.
4. Abe, R., Higashi, M. & Domen, K., 2010. Facile fabrication of an efficient oxynitride TaON photoanode for overall water splitting into H2 and O2 under visible light irradiation. *Journal of the American Chemical Society,* 132(34), pp. 11828-11829.

5. Abe, R., Higashi, M. & Domen, K., 2010. Facile fabrication of an efficient oxynitride TaON photoanode for overall water splitting into H2 and O2 under visible light irradiation. *Journal of American Chemical Society,* Volume 132, pp. 11828-11829.

6. adachi, M. et al., 2004. Highly Efficient Dye-sensitized solar cells with a titania thin-film electrode composed of a network structure of single-crystal-like TiO2 nanowires made by the "oriented attachement" mechanism. *Journal of The American Chemical Society,* 126(45), pp. 14943-14949.

7. Adesina, A. A., 1996. Hydrocarbon synthesis via Fischer-Tropsch reaction: travails and triumphs. *Applied Catalysis A: General,* 138(2), pp. 347-367.

8. Ager, J. W. et al., 2015. Experimental demonstrations of spontaneous solar-driven photoelectrochemical water splitting. *Energy and Environmental Science,* Volume 8, pp. 2811-2824.

9. Ahmed, S. & M.Krumpelt, 2001. Hydrogen from hydrocarbon fuels for fuel cells. *International Journal of Hydrogen Energy,* 26(4), pp. 291-301.

10. Alexander, B. D. et al., 2008. Metal oxide photoanodes for solar hydrogen production. *Journal of material Chemistry,* Volume 18, pp. 2298-2303.

11. Alivisatos, A., 1996. Semiconductor clusters, nanocrystals and quantum dots. *Science,* 271(5251), pp. 933-937.

12. Anikeeva, P. O., Halpert, J. E., Bawendi, M. G. & Bulovic, V., 2009. Quantum dot Light-emitting devices with electroluminescence tunable over the entire visible spectrum. *Nanoletters,* 9(7), pp. 2532-2536.

13. Anon., 1999. Chapter 1 - Introduction to PVD for microelectronics. *Thin Films,* Volume 26, pp. 1-21.

14. Arico, A. S. et al., 2017. Polymer Electrolyte Membranes for Water Photo-Electrolysis. *Membranes ,* 7(25), p. 020025.

15. Aroutiouniana, V. et al., 2006. Photoelectrochemistry of semiconductor electrodes made of solid solutions in the system Fe2O3-Nb2O5. *Solar Energy,* 80(9), pp. 1098-1111.

16. Aroutiounian, V., Arakelyan, V. & Shahnazaryan, G., 2002. Investigation of ceramic Fe2O3 (Ta) photoelectrodes for solar energy photoelectrochemical converters. *International Journal of Hydrogen Energy,* 27(1), pp. 33-38.
17. Aroutiounian, V., Arakelyan, V. & Shahnazaryan, G., 2005. Metal oxide photoelectrodes for hydrogen generation using solar radiation-driven water splitting. *Solar energy,* 78(5), pp. 581-592.
18. Awad, N. K., Ashour, E. A. & Allam, N. K., 2014. Recent advances in the use of metal oxide-based photocathodes for solar fuel production. *Journal of Renewable and Sustainable Energy,* Volume 6, p. 022702.
19. Bak, A., Choi, W. & Park, H., 2011. Enhancing the photoelectrochemical performance of hematite (a-Fe2O3) electrodes by cadmium incorporation. *Applied catalysis B: Environmental,* Volume 110, pp. 207-215.
20. Balkanski, M. & Wallis, R. F., 2000. *Semiconductor physics and applications.* 1st Edition ed. New York: Oxford University Press.
21. Banerjee, S., Mohapatra, S. K. & Misra, M., 2009. Synthesis of TaON nanotube arrays by sonoelectrochemcial anodization followed by nitridation: a novel catalyst for photoelectrochemical hydrogen generation from water. *Chemical Communications,* Volume 46, pp. 7137-7139.
22. Bard, A. J. et al., 1993. The electrode/electrolyte interface - A status report. *Journal of Physical Chemistry,* Volume 97, pp. 7147-7173.
23. Bar-Yam, Y. & Joannopoulos, J., 1987. Theories of defects in amorphous semiconductors. *Journal of non-crystalline solids,* Volume 97&98, pp. 467-474.
24. Berger, A., Segalman, R. A. & Newman, J., 2014. Material requirements for membrane separators in a water-splitting photoelectrochemical cell. *Energy and Environmental Science,* Volume 7, pp. 1468-1476.
25. Berglund, S. P. et al., 2016. Comprehensive evaluation of CuBi2O4 as a photocathode material for photoelectrochemical water splitting. *Chemistry of Materials,* Volume 28, pp. 4231-4232.
26. Berglund, S. P. et al., 2011. Photoelectrochemical oxidation of water using nanostructured BiVO4 films. *Journal of Physical Chemistry*

C, 115(9), pp. 3794-3802.

27. Bornoz, P. et al., 2014. A bismuth vanadate - cuprous oxide tandem cell for overall solar water splitting. *Journal of Physical Chemistry C,* 118(30), pp. 16959-16966.
28. Bossel, U., Eliasson, B. & Taylor, G., 2003. The future of hydrogen economy: brigth or bleak?. *Cogeneration and Competitive Power Journal,* 18(3), pp. 29-70.
29. Brillet, J. et al., 2010. Examining architectures of photoanode-photovoltaic tandem cells for solar water splitting. *Journal of Materials Research,* 25(1), pp. 17-24.
30. Brillet, J. et al., 2012. Highly efficient water splitting by a dual absorber. *Nature photonics,* Volume 6, pp. 824-828.
31. Cha, H. G. et al., 2010. Facile preparation of Fe2O3 thin film with photoelectrochemical properties. *Chemcial Communications,* 47(8), pp. 2441-2443.
32. Chatchai, P. et al., 2008. FTO/SnO2/BiVO4 composite photoelectrode for water oxidation under visible light irradiation. *Electrochemical and solid-state letters,* 11(6), pp. H160-163.
33. Chatchai, P. et al., 2009. Efficient photocatalytic activity of water oxidation over WO3/BiVO4 composite under visible light irradiation. *Electrochimica Acta,* 54(3), pp. 1147-1152.
34. Chen, H. et al., 2009. Progress in electrical energy storage system: A critical review. *Progress in natural Science,* Volume 19, pp. 291-312.
35. Chen, L., Dong, X., Wang, Y. & Xia, Y., 2016. Separating hydrogen and oxygen evolution in alkaline water electrolysis using nickel hydroxide. *Nature communications,* Volume 7, p. 11741.
36. Chen, Y. W. et al., 2011. Atomic layer-deposited tunnel oxide stabilizes silicon photoanodes for water oxidation. *Nature Materials,* Volume 10, pp. 539-544.
37. Chen, Z., Dinh, H. N. & Miller, E., 2013. *Photoelectrochemical Water Splitting.* Stanford, CA: Springer.
38. Chiang, C. Y., Shin, Y., Aroh, K. & Ehrman, S., 2012. Copper oxide photocathodes prepared by a solution based process. *Internationa; Journal of Hydrogen Energy,* Volume 37, pp. 8232-8239.

39. Cho, I. S. et al., 2015. Highly efficient solar water splitting from transferred TiO2 nanotube arrays. *Nano Letters,* 15(9), pp. 5709-5715.

40. Chuang, L. C., Moewe, M. & Chase, C., 2007. Critical diameter for III-V nanowires grown on lattice-mismatched substrates. *Applied Physics Letters,* Volume 90, p. 043115.

41. Cong, Y. et al., 2012. Tantalum Cobalt Nitride Photocatalysts for water oxiidation under visible light. *Chemistry of Materials,* 24(3), pp. 579-586.

42. Cong, Y. et al., 2012. Synthesis of Ta3N5 nanotube arrays modified with electrocatalysts for photoelectrochemical water oxidation. *Journal of Physical Chemistry C,* 116(27), pp. 14541-14550.

43. Coridan, R. H. et al., 2015. Methods for comparing the performance of energy conversion systems fo use in solar fuels and solar electricity generation. *Energy and Environmental Sciences,* Volume 8, pp. 2886-2901.

44. Cox, C. R., Lee, J. Z., Nocera, D. G. & Buonassisi, T., 2014. Ten percent solar-to-fuel conversion with nonprecious materials. *Proceedings of the National Academy of Sciences,* 111(39), pp. 14057-14061.

45. Dang, H. X. et al., 2012. Nanostructured Ta3N5 films as visible-light active photoanodes for water oxidation. *Journal of Physical Chemistry C,* 116(36), pp. 19225-19232.

46. Delahay, P. & Trachtenberg, I., 1956. Adsorption and Electrode Processes. *Journal of the American Chemical Society,* Volume 79, pp. 2355-2362.

47. Deng, J. et al., 2014. *Hematite nanostructures for high efficient solar water splitting.* s.l., IEEE Nanotechnology.

48. Deutsch, T., Koval, C. A. & Turner, J. A., 2006. III-V nitride epilayers for photoelectrochemcial water splitting: GaPN and GaAsPN. *Journal of Physical Chemistry B,* Volume 110, pp. 25297-25307.

49. Digdaya, I. et al., 2015. Extracting large photovoltages from a-SiC photocathodes with an amorphous TiO2 front surface field layer for solar hydrogen evolution. *Energy and Environmental Science,* Volume 8, pp. 1585-1593.

50. Dincer, I., 2000. Renewable energy and sustainable development: a crucial review. *Renewable and Sustainable Energy Reviews,* 4(2), pp. 157-175.

51. Doscher, H. et al., 2016. Solar-to-hydrogen efficiency: shining light on photoelectrochemcial device performance. *Energy and environmental science,* Volume 9, pp. 74-80.

52. Doyle, R. & Lyons, M., 2013. Kinetics and mechanistich aspects of the oxygen evolution reaction at hydrous iron oxide films in base. *Journal of the Electrochemical Society,* 160(2), pp. H142-H154.

53. Dry, M. E., 2002. The Fischer-Tropsch process: 1950-2000. *Catalysis Today,* 71(3-4), pp. 227-241.

54. Dunn, B., Kamath, H. & Tarascon, J.-M., 2011. Electrical Energy Storage for the Grid: A battery of choices. *Science,* 334(6058), pp. 928-935.

55. Edenhofer, O., Pichs-Madruga, R. & Sookona, Y., 2012. *Renewable Energy Sourcs and Climate Change Mitigation,* Postdam: IPCC.

56. EIA, 2016. *International energy Outlook 2016,* s.l.: EIA.

57. Ellabban, O., Abu-Rub, H. & Blaabjerg, F., 2014. Renewable energy resources: current status, future prospects and their enabling technology. *Renewable and Sustainable Energy Reviews,* Volume 39, pp. 748-764.

58. Faguy, P. W. et al., 1990. A study of bisulfate adsorption on Pt (111) single crystal electrodes using in situ Fourier transform infrared spectroscopy. *Journal of Electroanalytical Chemistry and Interfacial Electrochemistry,* 289(1-2), pp. 245-262.

59. Fang, Y.-L.et al., 2015. Advanced High accuracy scanning electron microscopy review methodology by virtual defect. *Joint Simpusion 2015 - eMDC and ISSM.*

60. Feng, J. S. X., Sloppy, J. D., Guo, L. & Grimes, C. A., 2011. Vertically aligned WO3 nanowires arrays grown directly on transparent consucting oxide coated glass: synthesis and photoelectrochemical properties. *Nano Letters,* Volume 11, pp. 203-208.

61. Feng, X. et al., 2010. Ta3N5 nanotube arrays for visible light water photoelectrolysis. *Nano Letters,* 10(3), pp. 948-952.

62. Fernando, C., Silva, L. D., Mehra, R. & Takahashi, K., 2001. Junction

effects of p-Cu2O photocathode with layers of hole transfer sites (Au) and electron transfer sites (NiO) at the electrolyte interface. *Semiconductor Science and Technology,* 16(6), pp. 433-439.

63. Fisher, G., Seacrist, M. R. & Standley, R. W., 2012. Silicon Crystal Growth and Wafer Technology. *Proceedings of the IEEE,* 100(Special Centenial Issue), pp. 1454-1474.
64. Floriano, E. A., Scalvi, L. V. d. A., Sambrano, J. R. & Geraldo, V., 2010. Evaluation of bulk and surfaces absorption edge energy of sol-gel-dip -coating SnO2 thin films. *Materials Research,* 13(4), pp. 437-443.
65. Fuji, K. et al., 2013. Characteristics of hydrogen generation from water splitting by polymer electrolyte electrochemical cell directly connected with concentrated photovoltaic cell. *International Journal of Hydrogen Energy,* 38(34), pp. 14424-14432.
66. Fujishima, A. & Honda, K., 1972. Electrochemical photolysis of water at a semiconductor electrode. *Nature,* Volume 238, pp. 37-38.
67. Fuoco, L., Joshi, U. A. & Maggard, P. A., 2012. Preparation and photoelectrochemical properties of p-type Cu5Ta11O30 and Cu3Ta7O19 semiconducting polycrystalline films. *Journal of Physichal Chemistry C,* 116(19), pp. 10490-10497.
68. Gaigalas, A. K., He, H.-J. & Wang, L., 2009. Measurement of Absorption and Scattering with an Integrating Sphere Detector. *Journal of Research of the National Institute of Standards and Technology,* 114(2), pp. 69-81.
69. Gaikwad, N. et al., 2005. Photoelectrochemical Characterization of Semitransparent WO3 films. *Journal of the Electrochemical Society,* 152(5), pp. G411-416.
70. Gaillard, N. et al., 2010. *Status of research on tungsten oxide-based photoelectrochemical devices at the University of Hawaii.* s.l., SPIE 7770, Solar Hydrogen and Nanotechnology.
71. Gamry instruments, n.d. *ntroduction to lectrochemical mpedance Spectroscopy.* [Online] Available at: https://www.gamry.com/assets/Uploads/Basics-of-Electrochemical-Impedance-Spectroscopy.pdf [Accessed 18 August 2017].

72. Gan, J., Lu, X. & Tong, Y., 2014. Towards highly efficient photoanodes: boosting sunlight-driven emiconductor nanomaterials for water oxidation. *Nanoscale,* Volume 6, pp. 7142-7164.
73. Gibson, T. L. & Kelly, N. A., 2008. Optimization of solar powered hydrogen production using photovoltaic electrolysis devices. *International Journal of Hydrogen Energy,* 33(21), pp. 5931-5940.
74. Glasscock, J. A., Barnes, P. R., Plumb, I. C. & Savvides, N., 2007. Enhancement of photoelectrochemical hydrogen production from hematite thin fims by the introduction of Ti and Si. *Journal of Physical Chemistry C,* 111(44), pp. 16477-16488.
75. Goodenough, J., 1971. Metallic oxides. *Progress in solid state chemistry,* Volume 5, p. 145.
76. Gori, M., Gruniger, H. & Calzaferri, G., 1980. Photochemical properties of sintered iron (III) oxide. *Journal of Applied Electrochemistry,* 10(3), pp. 345-349.
77. Gratzel, M., 2001. Photoelectrochemical cells. *Nature,* Volume 414, pp. 338-344.
78. Gratzel, M., 2003. Dye-sensitized solar cells. *Journal of Photochemistry and Photobiology C: Photochemistry reviews,* 4(2), pp. 145-153.
79. Gray, J., 2003. Ch3. In: *Handbook of Photovoltaic Science and Engineering.* s.l.:John Wiley and Sons.
80. Grimes, C. A. & Mor, G. K., 2009. TiO2 Nanotube Arrays: Application to Photoelectrochemcial Water Splitting. In: C. A. Grimes & G. K. Mor, eds. *TiO2 Nanotube Arrays: Synthesis, Properties and Applications.* s.l.:Springer, p. 149.
81. Gryse, R. D., Gomes, W. P., Cardon, F. & Vennik, J., 1975. On the interpretation of Mott-Schottky plots determined at semiconductor/ electrolyte systems. *Journal of the Electrochemical Society,* 122(5), pp. 711-712.
82. Guijarro, N., Prevot, M. S. & Sivula, K., 2015. Surface modification of semiconductor photoelectrodes. *Phys. Chem. CHem. Phys.,* Volume 17, pp. 15655-15674.
83. Hahn, N. & Mullins, B., 2010. Photoelectrochemical performance of nanostructured Ti and Sn-doped a-Fe2O3 photoanodes. *Chemistry of Materials,* 22(23), pp. 6474-6482.

84. Hahn, N. T., Holmberg, V. C., Korgel, B. A. & Mullins, C. B., 2012. electrochemical synthesis and characterization of p-CuBi2O4 thin film photocathodes. *ournal of Physical Chemistry C,* 116(10), pp. 6459-6466.

85. Han, L. et al., 2014. Efficient water-splitting device based on a bismuth vanadate photoanode and thin-film silicon soalr cells. *Chem. Sus. Chem.,* 7(10), pp. 2832-2838.

86. Han, L. et al., 2015. Gradient dopant profiling and spectral utilization of monolithic thin-film silicon photoelectrochemical tandem devices for solar water splitting. *Journal of Materials Chemistry A,* Volume 3, pp. 4155-4162.

87. Hernandez-Pagan, E. A. et al., 2012. Resistance and polarization losses in aqueous buffer-membrane electrolytes for water-splitting photoelectrochemcial cells. *Energy and Environmental Science,* Volume 5, pp. 7582-7589.

88. Hernandez-Pagan, E. A. et al., 2012. Resistance and polarization losses in aqueous buffer-membrane electrolytes for water-splitting photoelectrochemical cells. *Energy and Environmental Science,* Volume 5, pp. 7582-7589.

89. Heydenreich, J., 1993. Scancinc Electron Microscopy in Semiconductor Research. *Scanning,* Volume 15, p. 315.

90. Higashi, M., Domen, K. & Abe, R., 2011. Fabrication of efficient TaON and Ta3N5 photoanodes for water splitting under visible light irradiation. *energy and Environmental Science,* Issue 10.

91. Higashi, M., Domen, K. & Abe, R., 2011. Fabrication of efficient TaON and Ta3N5 photoanodes for water splitting under visible light irradiation. *Energy and Environmental Science,* Volume 4, pp. 4138-4147.

92. Higashi, M., Domen, K. & Abe, R., 2012. Highly stable water splitting on oxynitride TaON photoanode system under visible light irradiation. *Journal of the American Chemical Society,* 134(16), pp. 6968-6971.

93. Holloway, P. H. & McGuire, G. E., 1995. *Handbook of Compound Semiconductors: Growth, Processing, Characterization and Devices.* New Jersey: Noyes Publications.

94. Hong, S. J., Lee, S., Jang, J. S. & Lee, J. S., 2011. Heterojunction BiVO4/WO3 electrodes for enhanced photoactivity of water oxidation. *Energy and environmental science,* Volume 4, pp. 1781-1787.

95. Houlihan, J. et al., 1985. Substitutional and defect doping effects on hte photoelectrochemical properties of Fe2O3. *Materials Researh Bulletin,* 20(2), pp. 163-177.

96. Hu, S. et al., 2015. Thin-film materials for the protection of semiconducting photoelectrodes in solar fuel generators. *J. Phys. Chem C,* Volume 119, pp. 24201-24228.

97. Hu, S. et al., 2014. Amorphous TiO2 coatings stabilize Si, GaAs, and GaP photoanodes for efficient water oxidation. *Science,* 344(6187), pp. 1005-1009.

98. Hu, S. et al., 2013. An analysis of the optimal band gaps of light absorbers in integrated tandem photoelectrochemical water-splitting systems. *Energy and Environmental Science,* Volume 6, pp. 2984-2993.

99. Hu, Y.-S.et al., 2008. Pt-doped a-Fe2O3 thin films active for photoelectrochemical water splitting. *Chemistry of Materials,* 20(12), pp. 3803-3805.

100. Hybertsen, M. S. & Louie, S. G., 1986. Electron correlation in semiconductors and insulators: Band gaps and quasi particles. *Physical Review B,* 34(8), pp. 5390-5413.

101. Ida, S. et al., 2010. Preparation of p-type CaFe2O4 photocathodes for producting hydrogen from water. *Journal of American Chemical Society,* 132(49), pp. 17343-17345.

102. IEA, 2009. *A fundamental look at supply side enegry reserves for the planet,* s.l.: IEA.

103. Ishikawa, A. et al., 2004. Electrochemical behavior of thin Ta3N5 semiconductor film. *Journal of Physical Chemistry B,* 108(30), pp. 11049-11053.

104. Iwase, A. & Kudo, A., 2010. Photoelectrochemical water splitting using visible light responsive BiVO4 fine particles prepared in an aqueous acetic acid solution. *Journal of Maerials Chemistry,* 35(20), pp. 7536-7541.

105. Izumi, Y., 2013. Recent advances in the photoactalytic conversion of carbon dioxide to fuels with eater and/or hydrogen using solar energy and beyond. *Coordination Chemistry Reviews,* 257(1), pp. 171-186.

106. Jacobsson, J. T., Fjallstrom, V., Sahlberg, M. H. & Edoff, M., 2013. A monolithic device for solar water splitting based on series interconnected thin film absorbers reaching over 10 % solar-to-hydrogen efficiency. *Energy and Environmental Science,* 6(12), pp. 3676-3683.

107. Jang, J. S. et al., 2009. Rapid screening of effective dopants for Fe2O3 photocatalysts with scanning electrochemical microscopy and investigation of their photoelectrochemical properties. *Journal of Physical Chemistry C,* 113(16), pp. 6719-6724.

108. Jing, D. et al., 2010. Efficient solar hydrogen production by photocatalytic water splitting: from fundamental study to pilot demonstration. *International Journal of Hydrogen Energy,* 35(13), pp. 7087-7097.

109. Johnson, R. W., Hultqvist, A. & Bent, S. F., 2014. A brief review of atomic layer deposition: from fundamentals to applications. *Materials Today,* 17(5), pp. 236-246.

110. Jones, A. & Hitchman, M., 2009. *Chemical Vapour Deposition Precursors, Processes and Applicaitons.* s.l.:Royal Society of Chemistry.

111. Jong, P. E. d., Vanmakelbergh, D. & Kelly, J. J., 1999. cu2O: a catalyst for the photochemical decomposition of water?. *Chem Comm,* pp. 1069-1070.

112. Joshi, U. A., Palasyuk, A. M. & Maggard, P. A., 2011. Photoelectrochemical investigation and electonic structure of a p-type CuNbO3. *Journal of Physical Chemistry C,* 115(27), pp. 13534-13539.

113. Jo, W. et al., 2012. Phosphate doping into monoclinic BiVO4 for enhanced photoelectrochemical water oxidation activity. *Angew Chem Int Ed Engl,* 51(13), pp. 3147-3151.

114. Kainthla, R., Zelenay, B. & Bockris, J. O., 1987. Significant efficiency increase in self-driven photoelectrochemcial cell for water electrolysis. *Journal of the eletrochemcial Society,* 134(4),

pp. 841-845.

115. Kajari-Schroder, S., Kasewieter, J., Hensen, J. & Brendel, R., 2013. Lift-off of Free-standiing Layers in the Kerfless Porous Silicon Process. *Energy Procedia,* Volume 38, pp. 919-925.

116. Kaldellis, J. & Zafirakis, D., 2007. Optimum energy storage techniques for the improvement of renewable energy sources-based electricity generation economic efficiency. *Energy,* Volume 32, pp. 2295-2305.

117. Kargar, A. et al., 2013. #D branched nanowire photoelectrochemical electrodes for efficient solar water splitting. *ACS Nano,* 7(10), pp. 9407-9415.

118. Kargar, A. et al., 2013. Tailoring n-ZnO/p-Si branched nanowire heterostructures for selective photoelectrochemical water oxidation or reduction. *Nano Letters,* 13(7), pp. 3017-3022.

119. Kaunda, C. S., Kimambo, C. Z. & Nielsen, T. K., 2012. Hydropower in the context of sustainable energy supply: A review of technologies and challenges. *ISRN Renewable Energy,* Volume 2012, p. 730631.

120. Kawasaki, S. et al., 2012. Epitaxial Rh-doped SrTiO3 thin film photocathode for water splitting under visible light irradiation. *Applied Physics Letters,* 101(3), p. 033910.

121. Kay, A., Cesar, I. & Gratzel, M., 2006. New benchmark for water photooxidation by nanostructured a-Fe2O3 films. *Journal of the American Chemical Society,* 128(49), pp. 15714-15721.

122. Kay, A., Cesar, I. & Gratzel, M., 2006. New benchmark for water photooxidation by nanostructured a-Fe2O3 films. *Journal of the American Chemical Society,* 128(49), pp. 15714-15721.

123. Kennedy, J. H., Anderman, M. & Shinar, R., 1981. Photactivity of polycrystalline a-Fe2O3 electrodes doped with group IVA elements. *Journal of the electrochemical society,* 128(11), pp. 2371-2373.

124. Khaselev, O., Bansal, A. & Turner, J., 2001. High-efficiency integrated multijunction photovoltaic/electrolysis systems for hydrogen production. *International Journal of Hydrogen Energy,* 26(2), pp. 127-132.

125. Khaselev, O. & Turner, J., 1998. A monolithic photovoltaic-phototelectrochemical device for hydrogen production via water

splitting. *Science,* 280(5362), pp. 425-427.

126. Khaselev, O. & Turner, J. A., 1998. A monolithic photovoltaic-photoelectrochemical device for hydrogen production via water splitting. *Science,* 280(5362), pp. 425-427.

127. Khaselev, O. & Turner, J. A., 1999. Photoelectrolysis of HBr and HI usinga monolithic combined photoelectrochemical/photovoltaic device. *Electrochem. Solid-State Lett.,* Volume 2, pp. 310-312.

128. Khnayzer, R. S. et al., 2012. Structure and activity of photochemically deposited "CoPi" oxygen evolving catalyst on titania. *ACS Catalysis,* 2(10), pp. 2150-2160.

129. Kleiman-Shwarsctein, A. et al., 2008. Electrodeposition of a-Fe2O3 doped with Mo or Cr as photoanodes for photocatalytic water splitting. *Journal of Physical Chemistry C,* 112(40), pp. 15900-15907.

130. Kolodziej, A., 2004. Staebler-Wronski effect in amorphous silicon and its alloys. *Opto-electronics Review,* 12(1), pp. 21-32.

131. Kovalenko, M. V., 2015. Opportunities and challenges for quantum dot photovoltaics. *Nature Nanotechnology,* Volume 10, pp. 994-997.

132. Krol, R. v. d. & Gratzel, M., 2012. *Photoelectrochemical Hydrogen Production.* s.l.:Springer.

133. Kronawitter, C. X., Mao, S. S. & Antoun, B. R., 2011. Doped, porous iron oxide fims and their optical functions and anodic photocurrents for solar water splitting. *Applied Physics Letters,* Volume 98, p. 092108.

134. Kugler, S. & Shimakawa, K., 2015. *Amorphous semiconductors.* Cambridge: Cambridge University Press.

135. Kumari, S. et al., 2006. Characterization of Zn-dped hematite thin films for photoelectrochemical splitting of water. *Current Science,* 91(8), pp. 1062-1064.

136. Kumar, P. et al., 2011. Electrodeposited zirconium-doped a-Fe2O3 thin film for photoelectrochemical water splitting. *International Journal of hydrogen energy,* 36(4), pp. 2777-2784.

137. Kurobe, K.-i. et al., 2001. Effects of grain boundaries in polycrystalline silicon thin-film solar cells based on the two-dimensional model. *Solar energy materials & Solar cells,* Volume 65, pp. 201-209.

138. Law, D. C. et al., 2010. Future technology pathways of terrestrial III-V multijunction solar cells for concentrator photovoltaic systems. *Solar Energy Materials & Solar Cells,* Volume 94, pp. 1314-1318.

139. Lee, M. H. et al., 2012. p-type InP nanopillar photocathodes for efficient solar-driven hydrogen production. *Angewandte Chemie,* 51(43), pp. 10760-10764.

140. Lee, M. H. et al., 2012. p-type InP nanopillar photocathodes for efficient solar-driven hydrogen production. *Angew. Chem.,* Volume 51, pp. 10760-10764.

141. Lee, Y.-L., Chi, C.-F. & Liau, S.-Y., 2010. CdS/CeSe co-sensitized TiO2 photoelectrode for efficient hydrogen generation in a photoelectrochemical cell. *Chemistry of Materials,* 22(3), pp. 922-927.

142. Lemaire, P. & Gaspard, J., 1981. A simple description of the dangling bond state and H-passivation in amorphous semiconductors. *Solid State Communications,* 38(5), pp. 397-400.

143. Lettenmeier, P. et al., 2015. Nanosized IrOx-Ir catalyst with relevant activity for anodes of proton exchange membrane electrolysis produced by a cost-effective procedure. *Angewandte Chemie,* 55(2), pp. 742-746.

144. Lewerenz, H. J., Gerischer, H. & Lubke, M., 1984. Photoelectrochemistry of WSe2 electrodes: Comparison of Stepped and Smooth Surfaces. *Journal of The Electrochemical Society,* 131(1), pp. 100-104.

145. Liang, Y., Enache, C. S. & Krol, R. v. d., 2008. Photoelectrochemcial characterization of sprayed a-Fe2O3 thin films: influence of Si doping and SnO2 interfacial layer. *International Journal of Photoenergy,* Volume 2008, p. 739864.

146. Liang, Y., Tsubota, T., Mooij, L. P. & Krol, R. v. d., 2011. Highly Improved Quantum Efficiencies for thin film BiVO4 photoanodes. *Journal of Physical Chemistry C,* 115(35), pp. 17594-17598.

147. Liao, M. et al., 2012. Co3O4 nanoparticles as robust water oxidation catalysts towards remarkably enhanced photostability of a Ta3N5 photoanode. *Advanced Functional Materials,* 22(14), pp. 3066-3074.

148. Licht, S. et al., 2001. Over 18 % solar energy conversion to generation of hydrogen fuel; theory and experiment for efficient solar water splitting. *International Journal of Hydrogen Energy,* 26(7), pp. 653-659.

149. Li, L. et al., 2012. Visible light driven hydrogen production from a photo-active cathode based on a molecular catalyst and organic dye-sensitized p-type nanostructured NiO. *Chemical Communications,* Volume 48, pp. 980-990.

150. Lin, C.-Y., Lai, Y.-H., Mersch, D. & Reisner, E., 2012. Cu2O/NiOx nanocomposite as an inexpensive photocathode in photoelectrochemical water splitting. *Chemical Science,* Volume 3, pp. 3482-3487.

151. Ling, Y. et al., 2011. Sn-doped hematite nanostructures for photoelectrochemical water splitting. *Nano Letters,* 11(5), pp. 2119-2125.

152. Ling, Y. et al., 2011. Sn-doped hematite nanostructures for photoelectrochemical water splitting. *Nano Letters,* 11(5), pp. 2119-2125.

153. Linsbergler, A. L., Lu, G. & Yates Jr., J. T., 1995. Photocatalysis on TiO2 surfaces: Principles, mechanisms, and selected results. *Chemical Reviews,* Volume 95, pp. 735-758.

154. Lin, Y. et al., 2013. Amorphous Si thin film based photocathodes with high photovoltage for efficient hydrogen production. *Nano Letters,* 13(11), pp. 5615-5618.

155. Lin, Y. et al., 2015. role of TiO2 surface passivation on improving the performance of p-InP photocathodes. *Journal of Physical Chemistry C,* 119(5), pp. 2308-2313.

156. Lin, Y. et al., 2012. Growth of p-type hematite by atomic layer deposition and its utilization for improved solar water splitting. *Journal of the American Chemical Society,* 134(12), pp. 5508-5511.

157. Lin, Y. et al., 2011. Hematite-based solar water splitting: challenges and opportunities. *Energy and Environmental Science,* Volume 4, pp. 4863-4869.

158. Liu, C. et al., 2013. A fully integrated nanosystem of semiconductor nanowires for direct solar water splitting. *Nano Letters,* 13(6), pp. 2989-2992.

159. Liu, M. et al., 2013. Enhancing water splitting activity and chemcial stability of zinc oxide nanowire photoanodes with ultrathin titania shells. *Journal of Physical Chemistry C,* 117(26), pp. 13396-13402.

160. Liu, R., Zheng, Z., Spurgeon, J. & Yang, X., 2014. Enhanced photoelectrochemical water-splitting performance of semiconductors by surface passivation layers. *Energy and Environmental Science,* Volume 7, pp. 2504-2517.

161. Li, Y. et al., 2015. Efficient and stable photoelectrochemical seawater splitting with TiO2@g-C3N4 nanorod arrays decorated by Co-Pi. *The Journal of Physical Chemistry C,* Volume 119, pp. 20283-20292.

162. Li, Y. & Zhang, J. Z., 2010. Hydrogen generation from photoelectrochemical water splitting based on nanomaterials. *Laser Photonics Rev.,* 4(4), pp. 517-528.

163. Li, Y. et al., 2013. Cobalt phosphate-modified barium-doped tantalum nitride nanorod photoanode with 1.5 % solar energy conversion efficiency. *Nature communications,* Volume 4, p. 2566.

164. Li, Z. et al., 2013. Photoelectrochemcial cells for solar hydrogen production: current state of promising photoelectrodes, methods to improve their properties and outlook. *Energy and Environmental Science,* Volume 6, pp. 347-370.

165. Long, M., Cai, W. & Kirsch, H., 2008. Visible light induced photoelectrochemical properties of n-BiVO4 and n-BiVO4/p-Co3O4. *Journal of Physical Chamistry C,* 112(2), pp. 548-554.

166. Luo, H. et al., 2008. Structural and Photoelectrochemical properties of BiVO4 thin films. *Journal of Physical Chemistry C,* 112(15), pp. 6099-6102.

167. Luo, J. et al., 2014. Water photolysis at 12.3% efficiency via perovskite photovoltaics and Earth-abundant catalysts. *Science,* 345(6204), pp. 1593-1596.

168. Luo, J. et al., 2016. Bipolar membrane-assisted solar water splitting in optimal pH. *Advanced Energy Materials,* 6(13), p. 1600100.

169. Luo, W. et al., 2010. Synthesis, growth mechanism and photoelectrochemical properties of BiVO4 microcrystal electrodes. *Journal of Physics D: Applied Physics,* Volume 43, p. 405402.

170. Luo, W. et al., 2011. Solar hydrogen generation from seawater with a modified BiVO4 photoanode. *Energy and Environmental Science,* Volume 4, pp. 4046-4051.

171. Luo, W. et al., 2011. Solar hydrogen generation from seawater with a modified BiVO4 photoanode. *Energy and Environmental Science,* Volume 4, pp. 4046-4051.

172. Maeda, K. & Domen, N. N. K., 2009. a precursor route to prepare tantalum (V) nitride nanoparticles with enhanced photocatalytic activity for hydrogen evolution under visible light. *Applied Catalysis A: General,* 370(1-2), pp. 88-92.

173. Maeda, K. et al., 2011. SrNbO2N as a water-splitting photoanode with a wide visible-light absorption band. *Journal of the American Chemical Society,* 133(32), pp. 12334-12337.

174. Mao, Y. et al., 2012. Electrochemical synthesis of hierarchical Cu2O stars with enhanced photoelectrochemcials properties. *Electrochimica Acta,* Volume 62, pp. 1-7.

175. Mardani, A. et al., 2015. Sustainable and Renewable Energy: an overview of the application of multiple criteria decision making techniques and approaches. *Sustainability,* Volume 7, pp. 13947-13984.

176. Marsen, B., Cole, B. & Miller, E. L., 2008. Photoelectrolysis of water using thin copper gallium diselenide electrodes. *Solar Energy Materials and Solar Cells,* 92(9), pp. 1054-1058.

177. Maruthapandian, V. et al., 2017. Study of the oxygen evolution reaction catalytic behavior of CoxNi1-xFe2O4 in alkaline medium. *Applied Materials and Interfaces,* Volume 9, pp. 13132-13141.

178. Mase, K., Yoneda, M., Yamada, Y. & Fukuzumi, S., 2016. Seawater usable for production and consumption of hydrogen peroxide as a solar fuel. *Nature Communications,* Volume 7, p. 11470.

179. Matsumoto, Y., Omae, M., Sugiyama, K. & Sato, E., 1987. New photocathode materials for hydrogen evolution: calcium iron oxide (CaFe2O4) and strontium iron oxide (Sr7Fe10O22). *Journal of Physical Chemistry,* 91(3), pp. 577-581.

180. Matulionis, I. et al., 2008. *Development of a corrosion-resistant amorphous silicon carbide photoelectrodefor solar-to-hydrogen photovoltaic/photoelectrochemical devices.* s.l., Conference on

Solar Hydrogen and Nanotechnology III.

181. McCrone, A. et al., 206. *Global Trends in Renewable Energy Investment 2016,* s.l.: Frankfurt School (FS-UNEP).

182. Meissner, D. et al., 1988. Pinning and unpinning of band positions at the semiconductor/electrolyte interface. In: W. H. Bloss & F. Pfisterer, eds. *Advances in Solar Energy Technology.* s.l.:Elsevier, pp. 2999-3002.

183. Melania, M. W., Antonia, O. & Penev, M., 2013. *Blending Hydrogen into natural gas pipeline networks: A review of key issues,* s.l.: NREL.

184. Memming, R., 1998. *Semiconductor electrochemistry.* s.l.:Wiley.

185. Menyah, K. & Wolde-Rufael, Y., 2010. CO2 emissions, nuclear energy, renewable energy and economic growth in the US. *Energy Policy,* 38(6), pp. 2911-2915.

186. Mettee, H. & Calvin, J. W. O. M., 1981. Solar induced water splitting with p/n heterotype photochemical diodes: n-Fe2O3/p-GaP. *Solar Energy Materials,* 4(4), pp. 443-453.

187. Mettee, H., Otvos, J. & Calvin, M., 1981. Solar induced water splitting with p/n heterotype photochemical diodes: n-Fe2O3/p-GaP. *Sol. Energy Mater.,* Volume 4, p. 443.

188. Meyyappan, M., Delzeit, L., Cassell, A. & Hash, D., 2003. Carbon nanotube growth by PECVD: a review. *Plasma sources science and technology,* 12(2), pp. 205-216.

189. Miller, E. L. et al., 2010. Progress in new semiconductor materials classes of solar photoelectrolysis. *International JOurnal of Energy Research,* Volume 34, pp. 1215-1222.

190. Miller, E. L., Paluselli, D., Marsen, B. & Rocheleau, R. E., 2005. Development of reactively sputtered metal oxide films for hydrogen-producing hybrid multijunction photoelectrodes. *Solar Energy Materials and Solar Cells,* 88(2), pp. 131-144.

191. Miller, E. & Rocheleau, R., 2002. *Photoelectrochemical Production of Hydrogen.* Honolulu, Hawai, U.S. DOE Hydrogen Program Review.

192. Modestino, M. A. et al., 2014. Robust production of purified H2 in a stable, self regulating, and continuously operating solar fuel

generator. *energy and Environmental Science,* Volume 7, pp. 297-301.

193. Mor, G. K. et al., 2008. p-type Cu-Ti-O nanotube arrays and their use in self-biased heterojunction photoelectrochemical diodes for hydrogen generation. *Nano Letters,* 8(7), pp. 1906-1911.
194. Morisaki, H., Watanabe, T., Iwase, M. & Yazawa, K., 1976. Photoelectrolysis of water with TiO2-covered solar cell electrodes. *Applied Physiscs letters,* Volume 29, p. 338.
195. Neuman, J. & Thomas-Alyea, K. E., 2004. *Electrochemical Systems.* New Jersey: Wiley.
196. Ng, Y. H., Iwase, A., Kudo, A. & Amal, R., 2010. Reducing graphene oxide on a visible-light BiVO4 photocatalyst for an enhanced photoelectrochemical water splitting. *Journal of Physical Chemical Letters,* 1(17), pp. 2607-2612.
197. Nian, J., Hu, C. & Teng, H., 2008. Electrodeposited p-type Cu2O for H2 evolution from photoelectrolysis of water under visible light illumination. *International Journal of Hydrogen Energy,* Volume 33, pp. 2897-2903.
198. Nielander, A. C. et al., 2013. Photoelectrochemical behavior of n-type Si(111) electrodes coated with a single layer of graphene. *Journal of the American Chemical Society,* 135(46), pp. 17246-17249.
199. Nielander, A. C. et al., 2015. A taxonomy for solar fuels generators. *Energy and environmental Science,* Volume 8, pp. 16-25.
200. Ni, M., 2006. An Overview of Hydrogen Storage Technologies. *Energy Exploration and Exploitation,* 24(3), pp. 197-209.
201. Nozik, A. J., 1976. p-n photoelectrolysis cells. *Applied physics letters,* Volume 29, pp. 150-153.
202. Ohashi, K., McCann, J. & Bockris, J. O., 1977. Stable photoelectrochemical cell for the splitting of water. *Nature,* 266(5603), pp. 610-611.
203. Olah, G. A., 2005. Beyond Oil and Gas: The Methanol Economy. *Angewandte Chemie,* 44(18), pp. 2636-2639.
204. Oskam, G., Bergeron, B. V., Meyer, G. J. & C.Searson, P., 2001. Pseudohalogens for dye-sensitized TiO2 photoelectrochemical

cells. *Journal of Physical Chemistry,* 105(29), pp. 6667-6873.

205. Panella, B., Hirscher, M. & Roth, S., 2005. Hydrogen adsorption in different carbon nanostructures. *Carbon,* 43(10), pp. 2209-2214.
206. Paracchino, A. et al., 2011. Highly active oxide photocathode for photoelectrochemical water reduction. *Nature Materials,* Volume 10, pp. 456-461.
207. Paracchino, A. et al., 2011. Highly active oxide photocathode for potoelectrochemical water reduction. *Nature Materials,* 10(6), pp. 456-461.
208. Paracchino, A. et al., 2012. Ultrathin films on copper (I) oxide water splitting photocathodes: a study on performance and stability. *Energy and Environmental Science,* Volume 5, pp. 8673-8681.
209. Parent, L., Dodelet, J. & Dallaire, S., 1987. Plasma-Sprayed Semiconductor Electrodes. *Journal of Elecreochemical Society,* 134(9), pp. 2226-2233.
210. Park, H. S. et al., 2011. Factors in the metal doping of BiVO4 for improved photoelectrocatalytic activity as studied by scanning electrochemical microscopy and first-principles density-functional calculation. *Journal of Physical Chemistry C,* 115(36), pp. 17870-17879.
211. Park, H. S., Leonard, K. C. & BArd, A. J., 2013. Surface Interrogation Scannin Electrochemical Microscopy (SI-SECM) of photoellectrochemistry at a W/Mo-BiVO4 semiconductor electrode: quantification of hydroxil radicals during water oxidation. *Journal of Physichal Chemistry C,* 117(23), pp. 12093-12102.
212. Parkinson, B., 1984. On the efficiency and stability of photoelectrochemical devices. *Acc. Chem. Rev.,* Volume 17, pp. 431-437.
213. Park, J. H. & Bard, A. J., 2005. Unassisted water splitting from bipolar Pt/dye-sensitized TiO2 photoelectrode arrays. *Electrochemical and Solid-state Letters,* 8(12), pp. G371-G375.
214. Park, J. H., kim, S. & Bard, A. J., 2006. Novel carbon-doped TiO2 nanotube arrays with high aspect ratios for efficient solar water splitting. *Nano Letters,* 6(1), pp. 24-28.
215. Pavlishchuk, V. V. & Addison, A. W., 2000. Conversion constant for

redox potentials measured versus different reference electrodes in acetonitrile solutions at 25 C. *Inorganic Chimica Acta,* 298(1), pp. 97-102.

216. Pilli, S. K. et al., 2011. Cobalt-phosphate (Co-Pi) catalyst modified Mo-doped BiVO4 photoelectrodes for solar water oxidation. *energy and Environmental Science,* Volume 4, pp. 5028-5034.

217. Pinaud, B. A. et al., 2013. Technical and economic feasibility of centralized facilities for solar hydrogen production via photocatalysis and photoelectrochemistry. *Energy and Environmental Science,* Volume 6, pp. 1983-2002.

218. Qian, F., Wang, G. & Li, Y., 2010. Solar-driven microbial photoelectrochemical cells with a nanowire photocathode. *Nano Letters,* 10(11), pp. 4686-4691.

219. Raub, C. J., 1982. Electroplating of Palladium for Electrical Contacts. *Platinum Metals Rev.,* 26(4), p. 158.

220. Raval, M. C. & S.Solanki, C., 2013. Review of Ni-Cu Based Front Side Metallization for c-Si Solar Cells. *Journal of Solar Energy,* Volume 2013, p. 183812.

221. Read, C. G., Park, Y. & Choi, K.-S., 2012. Electrochemical synthesis of p-type CuFeO2 electrodes for use in a photoelectrochemical cell. *Journal of Physical Chemistry Letters,* 3(14), pp. 1872-1876.

222. Reece, S. et al., 2011. Wireless solar water splitting using silicon-based semiconductors and earth-abundant catalysts. *Science,* 4(334), pp. 645-648.

223. Reuter, S., Sousa, J. S., Stancu, G. D. & Helden, J.-P. H. v., 2015. Review on VUV to MIR absorption spectroscopy of atmospheric pressure plasma jets. *Plasma Sources Science and Technology,* Volume 24, p. 054001.

224. Rogelj, J. et al., 2016. Paris Agreement climate proposals need a boost to keep warming well below 2C. *Nature,* Volume 534, pp. 631-639.

225. Rovelli, L., Tilley, S. D. & Sivula, K., 2013. Optimization and stabilization of electrodepositied Cu2ZnSnS4 photocathodes for solar water reduction. *ACS Applid Materials and Interfaces,* 5(16), pp. 8018-8024.

226. Roxlo, C. B., Abeles, B. & Tiedje, T., 1994. Evidence for lattice-mismatch - induced defects in amorphous semiconductor heterojunctions. *Physical Review Letters,* 52(22).

227. Roy, S. C., Varghese, O. K., Paulose, M. & Grimes, C. A., 2010. Towards Solar Fuels: Photocatalytic Conversion of Carbon Dioxide to Hydrocarbons. *ACS Nano,* 4(3), pp. 1259-1278.

228. Saito, R., Miseki, Y. & Sayama, K., 2012. Highly efficient photoelectrochemical water splitting using a thin film photoanode of BiVO4/SnO2/WO3 multi-composite in a carbonate electrolyte. *Chem Comm,* Volume 48, pp. 3833-3835.

229. Sanchez, C., Sieber, K. & Somorjai, G., 1988. The photoelectrochemistry of nioium doped a-Fe2O3. *Journal of Electroanalytical Chemistry and Interfacial Electrochemistry,* 252(2), pp. 269-290.

230. Sartoretti, C. J. et al., 2005. Photoelectrochemical oxidation of water at transparent ferric oxide film electrodes. *Journal of Physical Chemistry B,* 109(28), pp. 13685-13692.

231. Satsangi, V. R. et al., 2008. Nanostructured hematite for photoelectrochemical generation of hydrogen. *International Journal of Hydrogen Energy,* 33(1), pp. 312-318.

232. Sayama, K. et al., 2006. Photoelectrochemical decomposition of water into H2 and O2 on porous BiVO4 thin-film electrodes under visible light and significant effect on Ag ion treatment. *J. Phys Chem B,* 110(23), pp. 11352-11360.

233. Sayama, K. et al., 2006. Photoelectrochemical decomposition of water into H2 and O2 on porous BiVO4 thin-fim electrodes under visible light and significant effect of Ag ion treatment. *Journal of Physical Chamistry B,* 110(23), pp. 11352-11360.

234. Sayama, K. et al., 2003. Photoelectrochemical decomposition of water of nanocrystalline BiVO4 film electrodes under visible light. *Chemical Communications,* Volume 23, pp. 2908-2909.

235. Sayama, K. et al., 2010. Effect of carbonates ions on the photooxidation of water over porous BiVO4 film photoelectrode under visible light. *Chemistry Letters,* 39(1), pp. 17-19.

236. Schneider, M., Schuller, I. K. & Rahman, A., 1987. Epitaxial growth of silicon: A molecular-dynamics simulation. *Physical Review B,*

Volume 36, p. 1340.

237. Schulz, H., 1999. Short history and present trends of Fischer-Tropsch synthesis. *Applied Catalysis A: General,* 186(1-2), pp. 3-12.

238. Schumacher, L. C., McIntyre, N. S., Mamiche-Afara, S. & Dignam, M. J., 1990. Photoelectrochemical properties of indium doped iron oxide. *Journal of Electroanalytical Chemistry and Interfacial Electrochemistry,* 277(1-2), pp. 121-138.

239. Schuttaufa, J.-W.et al., 2016. Solar-to-hydrogen production at 14.2 % efficiency with silicon photovoltaics and eart-abundant electrocatalysis. *Journal od the electrochemical society,* 163(10), pp. F1177-F1181.

240. Seabold, j. A. & Choi, k.-S., 2012. Efficient and stable photo-oxidation of water by a bismuth vanadate photoanode coupled with an iron oxyhydroxide oxygen evolution catalyst. *Journal of the American Chemical Society,* 134(4), pp. 2186-2192.

241. Seger, B. et al., 2013. Using TiO2 as a conductive protective layer for photocathodic H2 evolution. *Journal of the American Chemical Society,* 135(3), pp. 1057-1064.

242. Seshan, K., 2002. *Handbook of Thin film deposition Processes and Techiques.* Second ed. Santa Clara, California: Noyes Publications.

243. Shaner, M. R. et al., 2014. Photoelectrochemistry of core-shell tandem junction n-p+-Si/n-WO3 microwire array photoelectrodes. *Energy and Environmental Science,* Volume 7, pp. 779-790.

244. Shinar, R. & Kennedy, J. H., 1982. Photoactivity of doped a-Fe2O3 electrodes. *Solar Eneergy Materials,* 6(3), pp. 323-335.

245. Siripala, W. et al., 2003. A Cu2O/TiO2 heterojuntion thin film cathode for photoelectrocatalysis. *Solar Energy Materials and Solar Cells,* 77(3), pp. 229-237.

246. Sivula, K., 2012. Nanostructured a-Fe2O3 photoanodes. In: R. v. d. Krol & M. Gratzel, eds. *Photoelectrochemical hydrogen production.* new York: Springer.

247. Sivula, K. et al., 2010. Photoelectrochemical water spliiting with mesoporous hematite prepared by a solution-based colloidal approach. *Journal of the American Chemistry Society,* Volume 132, pp. 7436-7444.

248. Smets, A. et al., 2016. *Solar energy.* Cambridge: UIT Cambridge.

249. Spijker, H. v., Simon, D. & Ooms, F., 2008. Photocatalytic water splitting by means of undoped and doped La2CuO4 photocathodes. *International Journal of Hydrogen Energy,* 33(22), pp. 6414-6419.

250. Staebler, D. & Wronski, C. R., 1977. Reversible conductivity changes in discharge-produced amorphous Si. *Applied Physics Letters,* 31(4), pp. 292-294.

251. Stuke, J., 1970. Review of optical and electrical properties of amorphous semiconductors. *Journal of non-crystalline solids,* Volume 4, pp. 1-26.

252. Su, j., Guo, L., Bao, N. & Grimes, C. A., 2011. Nanostructured WO3/BiVO4 heterojunction films for efficient photoelectrochemical water splitting. *Nano Letters ,* 11(5), pp. 1928-1933.

253. Sun, K. et al., 2015. Sputtered NiOx films for stabilization of p+n-InP photoanodes for solar driven water oxidation. *Advanced Energy Materials,* Volume 5, p. 1402276.

254. Sun, K. et al., 2012. Metal on metal oxide nanowire Co-catalyzed Si photocathode for solar water splitting. *Nanotechnology,* 23(19), p. 194013.

255. Tachibana, Y., Vayssieres, L. & Durrant, J. R., 2012. Artificial photosynthesis for solar water-splitting. *Nature photonics,* Volume 6, pp. 511-518.

256. Tang, H. et al., 2012. Enhancing the stability of CuO thin-film photoelectrodes by Ti alloying. *Journal of Electronic Materials,* 41(11), pp. 3062-3067.

257. Tauc, J., 1968. Optical properties and electronic structure of amorphous Ge and Si. *Materials Research Bulletin,* 3(1), pp. 37-46.

258. Tomkiewicz, M. & Woodall, J. M., 1977. Photoelectrolysis of water with semiconductor materials. *Journal of the Electrochemical Society,* 124(9), pp. 1436-1440.

259. Tong, L. et al., 2012. Sustained solar hydrogen generation using a dye-sensitized NiO photocathode/BiVO4 tandem photo-electrochemical device. *Energy and Environmental Science,* Volume 5, pp. 9472-9475.

260. Trari, M., Bouguelia, A. & Bessekhouad, Y., 2006. p-type CuYO2

as a hydrogen photocathode. *Solar Energy Materials and Solar Cells,* 90(2), pp. 190-202.

261. Trindade, T., 2001. Nanocrystalline Semiconductors: Synthesis, Properties, and Perspectives. *Chemistry of Materials,* 13(11), pp. 3843-3858.
262. Trinidade, T., 2001. Nanocrystalline semiconductors: Synthesis, properties, and perspectives. *Chem. Mater.,* Volume 13, pp. 3843-3858.
263. Turan, B. et al., 2016. Upscaling of integrated phototelectrochemical water splitting devices to large areas. *Nature Communications,* Volume 7, p. 12681.
264. Ulmann, C. S. M. & Augustynski, J., 2001. Photoelectrochemical properties of nanostructured tungsten trioxide films. *Journal of Physical Chemistry B,* 105(5), pp. 936-940.
265. Uno, K., 1995. Energy demand and supply. In: *Environmental Options: Accounting for Sustainability.* Dodrecht, Nethelands: Springer, pp. 49-107.
266. Urbain, F. et al., 2016. Light-induced degradation of adapted quadruple junction thin film silicon solar cells for photoelectrochemical water splitting. *Solar Energy Materials and Solar Cells,* 145(2), pp. 142-147.
267. Urbain, F. et al., 2016. Multijunction Si photocathodes with tunable photovoltages from 2.0 V to 2,8 V for light induced water splitting. *Energy and Environmental Science,* Volume 9, pp. 145-154.
268. van de Krol, R. & Gratzel, M., 2012. *Photoelectrochemical Hydrogen Production.* s.l.:Springer.
269. Vasudev, M. C. et al., 2013. Exploration of Plasma-Enhanced Chemical Vapor Deposition as a Method for Thin-Film Fabrication with Biological Applications. *Applied Materials and Interfaces,* Volume 5, pp. 3983-3994.
270. Vernon-Parry, K., 2000. Scanning Electron Microscopy: an introduction. *III-V review,* 13(4), pp. 40-44.
271. Walter, M. G. et al., 2010. Solar Water Splitting Cells. *Chem. Rev.,* Volume 110, pp. 6446-6473.
272. Wang, D. et al., 2012. Photocatalytic water oxidation on BiVO4

with the elecrocatalyst as an oxidation cocatalyst: Essential relations between electrocatalyst and photocatalyst. *Journal of Physical Chemistry C,* 116(8), pp. 5082-5089.

273. Wang, D. et al., 2012. Photocatalytic water oxidation on BiVO4 with the electrocatalyst as an oxidation cocatalyst; Essential relations between electrocatalyst and photocatalyst. *Journal of Physcal Chemistry C,* 116(8), pp. 5082-5089.

274. Wang, L. et al., 2015. Anodic self-organized transparent nanotubular/pourous hematite film from Fe thin-films sputtered on FTO and photoelectrochemcial water splitting. *Res. Chem. Intermed.,* Volume 4, pp. 9333-9341.

275. Wang, X. et al., 2014. Silicon/hematite core/shell nanowire array decorated with gold nanoparticles for unbiased solar water oxidation. *Nano Letters,* 14(1), pp. 18-23.

276. Warren, S. C., 2011. Emergin Tredns in Water Photoelectrolysis. In: R. van de Krol & M. Gratzel, eds. *Photoelectrochemical Hydrogen Production.* Boston, MA: Springer, pp. 293-316.

277. Whiteman, A., Rinke, T., Esparrago, J. & Elsayed, S., 2016. *Renewable Capacity Statistics 2016,* s.l.: IRENA.

278. Xie, S., Zhang, Q., Liu, G. & Wang, Y., 2016. Photocatalytic and photoelectrochatalitic reduction of CO2 using heterogeneous catalysts with controlled nanostructures. *Chem. Commun.,* Volume 52, pp. 35-59.

279. Yacobi, B. & Holt, D., 1986. Cathodoluminescence scanning electron microscopy of semiconductors. *Journal of Applied Physics,* 59(4).

280. Yamada, Y. et al., 2003. One chip photovoltaic water electrolysis device. *International Journal of Hydrogen Energy,* 28(11), pp. 1167-1169.

281. Yang, J. et al., 2017. A multifunctional biphasic water splitting catalyst tailored for integration with high-performance semiconductor photoanodes. *Nature Materials,* Volume 16, pp. 335-341.

282. Ye, H., Lee, J., Jang, J. S. & Bard, A. J., 2010. Rapid screening of BiVO4-based photocatalysts by scanning electrochemical microscopy (SECM) and studies of their photoelectrochemical properties. *Journal of Physical Chemistry C,* 114(31), pp. 13322-

13328.

283. Yeh, T.-F., Teng, C.-Y., Chen, S.-J. & Teng, H., 2014. Nitrogen-doped graphene oxide quantum dots as photocatalysts for overall water-splitting under visible light illumination. *Advanced Materials,* 26(20), pp. 3297-3303.

284. Yndurain, F., 1992. Optical Properties of amorphous semiconductors. *Solid State Communications,* 84(1/2), pp. 217-220.

285. Yokoyama, D. et al., 2011. Ta3N5 photoanodes for water splitting prepared by sputtering. *Thin solid films,* 519(7), pp. 2087-2092.

286. Yoneyama, H., Sakamoto, H. & Tamura, H., 1975. A photo-electrochemical cell with production of hydrogen and oxygen by a cell reaction. *Electrochimica Acta,* 20(5), pp. 341-345.

287. Yoon, A. K.-Y., Noh, H. S. & Yoon, Y. S., 2016. Analysis of Vanadium Redox Flow Battery Cell with Superconducting Charging System for Solar Energy. *Electrical and Electronic Engineering,* 6(1), pp. 1-5.

288. Zeman, M., 2011. *Photovoltaic basics lecture notes.* Delft: TU Delft.

289. Zhang, M. et al., 2010. Improved photoelectrochemical responses of Si and Ti codoped a-Fe2O3 photoanode films. *Applied Physics Letters,* Volume 97, p. 042105.

290. Zhang, Z. et al., 2013. Carbon-layer-protected cuprous oxide nanowire arrays efficient water reduction. *ACS Nano,* 7(2), pp. 1709-1717.

291. Zhang, Z. & Wang, P., 2012. Highly stable coper oxide composite as an effective photocathode for water splitting via a facile electrochemical synthesis strategy. *Journal of Materials Chemistry,* Volume 22, pp. 2456-2464.

292. Zhang, Z. et al., 2013. Plasmonic gold nanocrystals coupled with photonic crystal seamlessly on TiO2 nanotube photoelectrodes for effiecnt visible light photoelectrochemical water splitting. *Nano Letters,* 13(1), pp. 14-20.

293. Zheng, J. Y., Song, G., Kim, C. W. & Kang, Y. S., 2012. Facile preparation of p-CuO and p-CuO/n-CuWO4 junction thin films and their photoelectrochemical properties. *Electrochimica Acta,* Volume 69, pp. 340-344.

294. Zhong, D. K., Choi, S. & Gamelin, D. R., 2011. Near-complete suppression of surface recombnation in solar photoelectrolysis by "Co-Pi" catalyst-modified W:BiVO4. *Journal of the American Chemical Society,* 133(45), pp. 18370-18377.

INDEX

T

V

W

X